Interactive Sports Technologies

Building on the unfolding and expanding embeddedness of digital technologies in all aspects of life, *Interactive Sports Technologies: Performance, Participation, Safety* focuses on the intersection of body movement, physical awareness, engineering, design, software, and hardware to capture emerging trends for enhancing sports and athletic activities. The accessible and inspiring compilation of theoretical, critical, and phenomenological approaches utilizes the domain of sports to extend our understanding of the nexus between somatic knowledge and human-computer interaction in general. Within this framework, the chapters in this volume draw upon a variety of concepts, processes, practices, and elucidative examples to bring together a timely assessment of interactive technologies' potential to facilitate increased performance, participation, and safety in sports.

The contents of the book are organized under five thematic areas—Embodiment, Content, Materiality, Information Architecture and Design, and Applications—which invite diverse perspectives from a wide range of academic and practice-based researchers within a comprehensive coverage of sport disciplines.

Veronika Tzankova is a PhD candidate in the School of Interactive Arts and Technologies, Simon Fraser University, Canada, and a Communications Instructor at Columbia College, Vancouver, Canada. Her background in human-computer interaction, communication, and sports shapes the essence of her research which explores the potential of interactive technologies to enhance physical awareness and motor-skill acquisition in high-risk sports activities.

Michael Filimowicz, PhD is Senior Lecturer in the School of Interactive Arts and Technology (SIAT) at Simon Fraser University, Canada. He has a background in computer mediated communications, audiovisual production, new media art, and creative writing. His research develops new multimodal display technologies and forms, exploring novel form factors across different application contexts including gaming, immersive exhibitions, and simulations.

Interactive Sports Technologies

Performance, Participation, Safety

**Edited by Veronika Tzankova and
Michael Filimowicz**

Routledge
Taylor & Francis Group

NEW YORK AND LONDON

First published 2022
by Routledge
605 Third Avenue, New York, NY 10158

and by Routledge
4 Park Square, Milton Park, Abingdon, Oxon, OX14 4RN

*Routledge is an imprint of the Taylor & Francis Group, an informa
business*

© 2022 Taylor & Francis

Library of Congress Cataloging-in-Publication Data
Names: Tzankova, Veronika, editor. | Filimowicz, Michael, editor.
Title: Interactive sports technologies : performance, participation,
safety / edited by Veronika Tzankova and Michael Filimowicz.
Description: New York, NY : Routledge, 2022. | Includes
bibliographical references and index.
Identifiers: LCCN 2021061848 (print) | LCCN 2021061849 (ebook) |
ISBN 9780367506094 (hardback) | ISBN 9781032070490 (paperback) |
ISBN 9781003205111 (ebook)
Subjects: LCSH: Sports--Technological innovations. | Physical
education and training--Technological innovations. | Performance
technology. | Sports--Safety measures. | Human-computer interaction.
Classification: LCC GV745 .I67 2022 (print) | LCC GV745 (ebook) |
DDC 796.028/4--dc23/eng/20220126
LC record available at https://lccn.loc.gov/2021061848
LC ebook record available at https://lccn.loc.gov/2021061849

ISBN: 978-0-367-50609-4 (hbk)
ISBN: 978-1-032-07049-0 (pbk)
ISBN: 978-1-003-20511-1 (ebk)

DOI: 10.4324/9781003205111

Typeset in Times New Roman
by MPS Limited, Dehradun

Contents

Figures

Tables

Contributors

Josh Andres is a researcher and designer at the Exertion Games Lab at Monash University, Australia whose goal is to investigate and design intelligent systems as partners to enable human potential. He holds a PhD from RMIT University, where he investigated extending humans' cognitive and physical abilities through a human-computer integration approach. His Twitter handle is @ExperiencePlay

Scott Boatright is a cognitive psychologist and instructional designer. Currently, he is a PhD Student in Learning Systems Design & Technology at SIU. He has a B.S. in Psychology and a M.A. in Applied Experimental Psychology. His research interests include high-speed decision-making, deliberate practice, expertise, instructional design, and interactive learning technologies. Previously, he has contributed to the development of training technologies for the military and baseball umpires, as well as trained K–12 students on perceptual-cognitive skills to assist with learning.

Professor Chris Button is the Dean of the School of Physical Education, Sport and Exercise Sciences. He gained his PhD in Motor Learning at Manchester Metropolitan University in 2000 and then worked at Edinburgh University before moving to Otago in 2003. The processes by which humans learn to move skilfully is the central theme for Chris' research. Chris is passionate about movement and he uses models from sport and physical activity to explain skill acquisition. His research outputs cover a broad range of topics including the assessment of movement competence in children, individual differences, and sports technology. The theory of ecological dynamics, which underpins Chris' research, is featured in several textbooks he has co-authored. Over the last ten years, Chris has been working closely with Water Safety New Zealand to improve the learning of water safety skills amongst New Zealanders.

Kenneth Cortsen, MSc in business economics, MBA in general management and PhD with emphasis on the capitalization on sports branding. He consults and does research in various areas of commercialization of sports and is a Co-Founder of the Sport Management program at University College of Northern Denmark (UCN) in Denmark, which was initiated in 2009. In addition to its ordinary on-campus activities, the program also offers online sport management education to professional athletes (many of them are footballers) across Europe in collaboration with the World Players' Union (FIFPro) in Amsterdam. He also teaches and does research across the world, including guest lectures or research visits at University of San Francisco, Vlerick Business School, Harvard Business School, University of Northern Colorado, DIS Copenhagen, and Johan Cruyff Institute (Amsterdam and Barcelona). Since 2014, Kenneth, who holds a UEFA A-license, has worked as an elite football coach in the Danish top club Aalborg BK/AaB in various positions, e.g., head coach for men's reserves, head coach for women's team, which he took from the third tier to the best league and now he is a strategic advisor for the board of the mother club. In addition, Cortsen has been a part of various committees under the Danish FA and in the Danish league system while being an advisory board member of the IT-company, KMD's focus on data in sports. He is also Executive Board Member & Football Business Expert for the international football consulting agency Four Nations Football Consulting in Barcelona.

Dr. Peter Fadde is a professor and director of the Learning Systems Design & Technology graduate program at Southern Illinois University, USA. Previously, he served for 13 seasons as athletics video coordinator at Purdue University, where he earned masters and doctoral degrees. Dr. Fadde originated the Expertise-Based Training (XBT) method that repurposes expertise research methods into expertise training methods. Dr. Fadde was named SIU's Innovator-of-the-Year in 2013 and holds a patent on computer software for tennis stroke recognition.

Dr. P. David Howe holds the Dr. Frank J. Hayden Endowed Chair in Sport and Social Impact in the School of Kinesiology at Western University, Canada. David's ethnographic research focuses on unpacking the embodied socio-cultural milieu surrounding inclusive physical activity and disability sport. David is also editor of the Routledge Book Series, *Disability, Sport and Physical Activity Cultures* and holds a guest professorship at Katholieke Universiteit Leuven, Belgium.

Tobias Langlotz is Associate Professor in the University of Otago, New Zealand. Tobias was previously a senior researcher at the Institute for Computer Graphics and Vision (Graz University of Technology, Austria) where he also obtained his PhD. Tobias' main research interest is Vision Augmentations and Computational Glasses utilizing AR technology, spontaneous interaction for wearable AR systems, and nomadic mobile telepresence solutions, where he works at the intersection of HCI, Computer Graphics, Computer Vision and Ubiquitous Computing.

Wei Hong Lo is currently a PhD student at the University of Otago, New Zealand majoring in the field of Augmented Reality Visualization and Human-Computer Interaction (HCI). Before his PhD studies, he graduated with a BSc (Hons.) Software Engineering in the University of Nottingham Malaysia Campus, where he obtained an interest in the field of HCI and visualization. He then proceeded as a research assistant and was involved in projects such as developing interactive exhibits for museums and exhibitions before starting his PhD.

Daniel Almeida Marinho finished his PhD in Sport Sciences at the University of Trás-os-Montes and Alto Douro (UTAD, Vila Real, Portugal). Currently he holds a position of Associate Professor with tenure at the UBI, for teaching and researching Swimming and Biomechanics Analysis in different Sports and Activities. He is the vice-director from the Research Center for Sports, Health and Human Development (CIDESD). He has published more than 200 original articles in peer-reviewed journals and have been supervisor of more than 10 PhD and approximately 50 Masters students.

Mário Cardoso Marques is the Head and full Professor of Training Theory and Methodology in the Department of Sports Sciences at the University of Beira Interior, Portugal. He also holds an appointment as a Head of Strength and Conditioning performance analysis team of the CIDESD. Collaborator in several funded projects focused on frailty and exercise to promote and recommend changes in lifestyle associated with physical exercise for frail patients at risk of functional decline and MID-FRAIL. He has published more than 200 original articles in peer-reviewed journals and have been supervisor for more than 10 PhD and approximately 40 Masters students.

Diogo Luís Marques holds a BSc and an MSc degree in Sport Sciences from the University of Beira Interior (UBI). He is currently a research fellow from the Portuguese Foundation for Science and Technology (FCT), a PhD student in UBI, and a collaborator in the CIDESD. He analysis how movement velocity can be used during resistance training to assess, prescribe, and monitor the training load in young and older adults.

Steven Mills is an Associate Professor at the University of Otago, New Zealand where he gained his PhD in 2000. Between being a student and an academic at Otago he worked in a variety of commercial research and development roles and as a lecturer in The University of Nottingham. His interests lie in computer vision, particularly 3D reconstruction from images and applications with cultural and heritage value.

Florian 'Floyd' Mueller is Professor in the department of Human-Centred Computing of Monash University in Melbourne, Australia, directing the Exertion Games Lab that investigates the coming together of technology, the human body and play. Previously, he was in RMIT, Stanford, University of Melbourne, Microsoft Research, MIT Media Lab, Fuji-Xerox Palo Alto Labs, Xerox Parc, and the Australian CSIRO. Floyd was appointed to be general co-chair for CHI 2020, the ACM SIGCHI Conference on Human Factors in Computing Systems, the premier publication outlet for the Human-Computer Interaction (HCI) discipline.

Henrique Pereira Neiva concluded his PhD in Sport Sciences at the UBI in 2015. Currently she holds a position of Assistant Professor at the UBI. Member of the CIDESD since 2015. In the last years, produced more than 100 scientific documents and reviewed several peer-reviews papers in high-ranked journals. The current area of research is training control and evaluation, specifically what concerns to warm-up and performance. Recently he started to research on strength training and concurrent training.

Emily Nicol is a current PhD candidate at Griffith University, Australia. She has been involved in high performance sport for the past ten years as both an athlete and academic. Emily's current areas of research interest include individualized velocity profiles, anthropometric characteristics and timing patterns in elite swimmers. Emily also has a strong interest in the practical application of research outcomes to the daily training environment. Emily's research has been presented at a number of international conferences and published in well-regarded academic journals.

Takuya Nojima received a PhD in Engineering from The University of Tokyo, Tokyo, Japan, in 2003. He is currently Associate Professor in the Department of Informatics University of Electro-Communications, Tokyo, Japan. His research interests include haptic interaction, superhuman sports, human interface, and virtual reality.

Daniel Rascher teaches and publishes research on sports business topics and consults to the sports and entertainment industry. He specializes in economics and finance and more specifically in industrial organization, antitrust, competition analysis, M&As, valuation, analytics, econometrics, economic impact, market readiness, feasibility research, marketing research, damages analysis, class certification, strategy, and labor issues in the sports industry. Dr. Rascher founded SportsEconomics to enable sports enterprises to capitalize on the sports industry's transition from hobby status to multi-billion dollar industry. As Founder and President of SportsEconomics, LLC, Partner at OSKR, LLC, and former Principal at LECG, LLC, his clients in more than 150 engagements have included organizations involved in the NBA, NFL, MLB, NHL, NCAA, NASCAR, MLS, PGA, media, sporting goods and apparel, professional boxing, mixed martial arts, minor league baseball, NHRA, AHL, F1 racing, Indy Car racing, American Le Mans racing, Premier League

Football, NASL, women's professional soccer, professional cycling, endurance sports, Indian Premier League, ticketing, IHRSA, youth sports, music, as well as the FTC, IRS, EEOC, sports commissions, local and state governments, convention and visitors bureaus, tourism businesses, entrepreneurs, and B2B enterprises.

Kadri Rebane is a Doctorate student at the University of Electro-Communications in Tokyo, Japan. She belongs to the augmented sports research group in the computer-human interfaces lab led by Takuya Nojima. Her research interests are augmented sports and human interface.

Dr. Holger Regenbrecht is a computer scientist and Professor at the Department of Information Science at the University of Otago, New Zealand. He received his doctoral degree Dr. Ing. from Bauhaus University Weimar, Germany. Holger has been working in the fields of virtual and mixed reality for more than 20 years. His work spans theory, concepts, techniques, technologies, and applications. Holger's research interests include human-computer interaction, applied computer science and information technology, presence and telepresence, ubiquitous, immersive, and collaborative augmented reality, three-dimensional user interfaces, psychological aspects of virtual and mixed reality, computer-aided therapy, and VR learning and collaboration.

Nina Schaffert is an associated research fellow in Human Movement Science and received her PhD from the University of Hamburg. Her research areas include Movement Sonification; Measuring and Feedback Technologies; Motor Learning and Control; and Movement Analysis. She applies this expertise to research developing applications for sports practice including auditory displays; investigating methods of improving user's performance; and investigating how sonification designs need to be applied for use in sport and rehabilitation in cooperation with industrial partners. She also works as scientific coordinator at BeSB GmbH Berlin.

Sebastian Schlüter graduated from the Technical University of Berlin with a Master's degree in Audio Communication and Technology. He is currently employed as an acoustic engineer in the BeSB GmbH Berlin and is involved in several research and development projects with the Department of Human Movement Science of the University of Hamburg. He applies his expertise to develop movement analysis systems based on state-of-the-art sensor technology with specially adapted sonification algorithms and real-time sound synthesis.

Dr. Carla Filomena Silva is an assistant professor in the School of Health Studies, at Western University, Canada. Before starting to work in Western University, Carla has worked at Nottingham Trent University, UK, and for the UNESCO chair in Inclusive PE, Sport, Fitness and

Recreation, in the Institute of Technology, Tralee, Ireland. Carla's eclectic research interests are situated at the nexus of sport, health, marginal bodies, and social justice.

Alice Sweeting has been a Research Fellow with Victoria University, Australia since 2016, after completing her PhD with Netball Australia and the Australian Institute of Sport. Alice currently supervises PhD students, as part of a strategic partnership between Victoria University and the Western Bulldogs Football Club. These sports analytics projects include applying data mining techniques to wearable sensor and skilled output data, to evaluate team-sport matches and training. Alice's research has been published in high-quality journals and presented at international conferences. Alice's key areas of interest include spatiotemporal data analysis, the use of wearable sensors and the complex systems approach to understanding team-sport behaviour. Alice is also passionate about programming in R and currently teaches into postgraduate sports analytics courses at Victoria University.

Elaine Tor has completed a PhD with Victoria University, Swimming Australia and the Australian Institute of Sport and been an Applied Sport Biomechanist and Researcher for over ten years. Elaine has worked directly with athletes in multiple sports including swimming, athletics, shooting, triathlon and archery. Elaine's work largely centres around utilizing data and technology to assist athletes and coaches to find competitive advantages to improve their performances at the highest level. Elaine has authored multiple peer-reviewed journal articles, presented at internationals conferences and written book chapters. Her key areas of research interest are swimming biomechanics, interactive data visualizations and using biomechanics to individualize technique improvement. Elaine has worked as the Lead Biomechanist at the Victorian Institute of Sport and holds an Adjunct Research Fellow Position at Victoria University.

Stefanie Zollmann is Senior Lecturer at the University of Otago, New Zealand. Before, she worked at Animation Research Ltd on XR visualization and tracking technology for sports broadcasting and urban planning. She worked as a post-doctoral researcher in the Institute for Computer Graphics and Vision (Graz University of Technology) where she also obtained a PhD degree in 2013. Her main research interests are XR for sports and media, visualization techniques for augmented reality, but also include capturing for XR and immersive experiences.

1 Introduction: A Comprehensive Approach to Interactive Sports Technologies

Veronika Tzankova and Michael Filimowicz

1.1 Introduction

Despite the widespread interest in digital and interactive technologies for exercise and movement—as evidenced by a variety of commercial products, patents, and publications—there is a surprising absence of methodically organized academic research on the use of interactive technologies within sports and sports-related activities. Given that sports are one of the most accessible form of systematized human movement, paired with the continuously increasing ubiquity of computation within everyday life, the goal of *Interactive Sports Technologies* is to outline the thematic and practical issues underlying the relationship between computing, human-computer interaction (HCI), embodiment, movement + athletic skill, and interactive technologies for sports. Specifically, the anthology takes into account the continuously expanding horizons of HCI to situate the research, methodologies, design, utilization, and experience of interactive technologies as central to the acquisition of somatic knowledge, awareness, feel, perception, self-imaging, and motivation—among others—directed toward increased athletic performance in sports. At the same time, it is also designed as an exploration into the alternative possibilities that fuse the productive and imaginative potential of embodied-sports practices with HCI research and digital technologies. It intends to provide a context, inspiration, and structure to facilitate new forms of inquiry into the emergent domain of interactive sports technologies (Figure 1.1).

At the same time, it is also designed as an exploration into the alternative possibilities that fuse the productive and imaginative potential of embodied-sports practices with HCI research and interactive technologies. It intends to provide context, inspiration, and structure to facilitate new forms of inquiry into the emergent domain of interactive sports technologies.

1.2 The Challenges Surrounding the Research + Design of Interactive Sports Technologies

One of the primary challenges surrounding the research and design of interactive sports technologies has been the identification of a 'common ground'

DOI: 10.4324/9781003205111-1

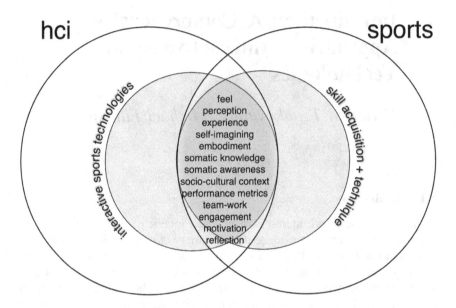

Figure 1.1 Explorations at the verge of movement expertise, HCI, and interactive technologies for sports.

that most efficiently captures the complexity and totality of relations involved in these. The range of thematic domains implicated are numerous, including HCI, embodiment theory + practice, athletic skill development, research methodologies, design approaches, data implementation, and the varying specifics of sport disciplines. Taken individually, each of these thematic domains comes associated with a wide range of practical, conceptual, and methodological traditions. When viewed in aggregate, the challenge becomes identifying points of departure that collectively plot the details and nuances formed by the intricate interactions and interrelations of the aforementioned fields without trivializing the variations that may emerge from adopting a specific perspective.

Currently, a substantial part of the research into interactive sports technologies comes from the field of HCI. A review of the available literature within this domain reveals several thematic tensions that condition 'unilateral' approaches which inform the theory and practice of these. Although such approaches address and solve many practical problems, they limit the possibilities of expanding and reimagining the potential of interactive technologies for sports. Drawing on the historical influences, values, and assumptions that drive the field of interactive sports technologies within HCI, we have identified three main thematic tensions that we discuss in detail in the following sections of this introductory chapter: (1) between efficiency and capacity for experience; (2) mind-body dualism; and (3) goal-oriented vs. altruistic enframings of technology.

1.2.1 Between Efficiency and Capacity for Experience

The introduction of embodiment theory and practice to the domain of HCI has created significant opportunities for rethinking the role technologies play in our lives. The intellectual tradition associated with the study of technology has emphasized that technology tends to operate by its own inherent logic of efficiency (see e.g. Ellul, 1973; Feenberg, 2010; Heidegger, 1996; Marx & Engels, 1983). Technology becomes a product of its function stripped bare of social contexts or cultural meanings. The function of technology dictates its trajectory of development where the logic of efficiency serves as a basis for identifying that trajectory. Thus, technologies are inherently subordinated and operate by the logic of efficiency. Efficiency is, as Heidegger (1996) suggests, an aspiration to control nature and being in itself.

Although philosophers of technology have situated efficiency more as a theoretical construct that helps explain the immanent striving for technological progress in the face of environmental and social catastrophes, we can trace its practical dimension reflected in the first and second wave HCI. The first wave HCI "was concerned with engineering systems to make people working with machines more effective" (Cairns & Power, 2018, p. 61), where the second wave extended on the human-technology nexus to involve collaborative, mediated, and distributed applications within work settings (Bødker, 2015). Both of these waves reflect on the preoccupation with assisting people to be faster and make fewer mistakes—an overt example of the practical modality of efficiency. Third wave HCI, with its strong focus on experience, has transformed the essence of human-computer interaction to engage with values, meaning-making, and situated knowledge (ibid). Body-based approaches to interactive technology design fall into the third wave and seem to have further emphasized the importance of lived experiences and felt senses to technological mediation. We can say that third wave HCI shifted—at least partially—the perceived purpose of technology from efficiency to facilitation of experience. Based on this trajectory of development, we can argue that the design and configuration of contemporary technologies should do more than merely accomplish our ends; through enhancing a space for reflection, technologies carry the potential to remind us of the ameliorative experiential properties of meaningful relations with both ourselves and the world outside us.

Where approaching the design of interactive technologies with a specific focus on lived experiences has already been implemented (see e.g. Antle et al., 2011; Cairns & Power, 2018; Hallnäs & Redström, 2001), the domain of interactive sports technologies seems to predominantly be operating under the presumption of efficiency (Nylander et al., 2014). This forms a thematic tension, where athletic skills are treated as a measurable performance outcome without consideration of non-measurable somatic self-mastery and "skills of experience" (Loke & Schiphorst, 2018, para. 4) which are in many cases essential to athletic operations.

We are currently experiencing a new wave of digital transformation, where smart materials, autonomous technologies, and ubiquitous systems shift the way in which we navigate through our lives and experiences. Kristina Höök argues that these changes present a significant opportunity to reimagine "the way we interact with the inanimate world" (2018, p. 2), where designing engaging interactions with technologies is important. The design of engaging interactions heavily depends on establishing a positive flow of experience between the soma and the technology where somaesthetic sensibilities are essential and should be given substantial consideration. Reimagining the potential of interactive sports technologies can help us extend on Höök's insight by considering the ways in which accounting for somaesthetic sensibilities in technology [design + interaction + application] can mediate our relations not only with the inanimate world, but also our relations with ourselves and the rest of the animate world.

This insight appeared to the first author of this chapter during one of her horseback riding sessions. She recollects:

> *My horse seemed to object to everything I did. Although I was closely following the instructions of my trainer, I was consistently failing to establish a positive relational flow with my equine partner. At some point, my trainer made a remark that stuck with me for its insightful simplicity. She said: "Maybe you should first learn to control yourself before attempting to influence anything else." The trainer's remark made me shift attention from my horse (and the way he was doing everything 'wrong') to myself, and my own state of body/mind. This attentional shift fixed all problems.*

Without realizing it, the trainer got engaged in the centuries-long philosophical quest for self-mastery, which has been an integral part of ethics and aesthetics in their pursuit of Socrates' question "What is a good life?" The art of equitation—where self-mastery establishes the 'ground zero'—is the perfect example of how "the joys and simulations of so-called pure thought are (for us embodied humans) influenced by somatic conditioning"(Shusterman, 2008, p. 21). Sometimes we need external input to navigate our state of body/mind in the right direction and emphasize our somatic potentials. Often such input comes from humans, but this is not a necessity. We can create technologies that support ameliorative shifts of attention, stimulate somatic awareness, and assist the users to reflect on themselves.

As an equestrian rider, Höök has also observed the parallels between human-horse coupling and human-machine coupling. In her article *Transferring Qualities from Horseback Riding to Design* (2010), Höök synthesizes the lessons extracted from learning horseback riding—specifically, awareness of body balance, weight, muscle memory, flow and rhythm of movement and how such lessons can be transferred to design principles. Höök suggests that these can be utilized by treating (full-) body movement as (1) wordless signs and signals; (2) bodily learning; and (3) aesthetic experiences through rhythm. The unification of different agents within the construction of an assembly or a closed system through non-verbal communication is similar to treating human-computer coupling as a synergetic, co-evolving relation. The aesthetic experience is created in the vitality of flow as an all-absorbing state. Movement (riding in this case) can facilitate aesthetic experiences and a certain quality of interaction as rhythm is often essential to body functions & movements (heartbeat and breathing, for example). Rhythm can make the "interaction come alive" (ibid, p. 234).

Höök's article corresponds to the emergent consideration of somatics practices in HCI. Somatics is defined as "the field which studies the *soma*: namely the body as perceived from within by first-person perception" (Hanna, 1986, p. 4). Richard Shusterman (2008) extends on the conceptualization of the 'soma' to develop a systematic philosophical framework called somaesthetics. Somaesthetics is concerned with "the critical study and meliorative cultivation of how we experience and use the living body (or soma) as a site of sensory appreciation (aesthesis) and creative self-fashioning" (ibid, p.19). Somaesthetics has since been adapted into the field of HCI as an experience-centered approach for exploring the aesthetic dimensions of interaction and improving designers' sensibilities of haptic, dynamic, and non-material qualities of movement. Current adaptations of somaesthetics in HCI predominantly focus on borrowing its theoretical instruments to explain human movement, where somaesthetics' pragmatic and practical branches—developing specific methods for somatic improvement and actual implementation of somatic improvement, respectively—have so far been overlooked (Lee et al., 2014).

Nevertheless, the turn to experience (Dourish, 2001) in HCI along with a systematic rediscovery of the body as a place of knowledge production (Varela et al., 1991) and aesthetic appreciation (Dewey, 1997; Shusterman, 2008) in the Western intellectual tradition have triggered the reconsideration of the historically-dominant understanding of human-technology relations. The idea of technological efficiency has been challenged and slowly shifted toward what can be summarized as elevated sociotechnical conscience, where somatics and somatic values constitute a major realm of consideration.

With an aging, overweight population and many lifestyle-associated diseases on the rise, sports have been garnering wider attention and prioritization in public policy and research. This has initiated a more generous distribution of resources to academic inquiry into sports across many disciplines, where HCI is one of the fields fostering such inquiry. Third-wave HCI and the distinctive phenomenological dimension associated with it have placed the notion of experience as central to our understanding of the usability and integrity of digital technologies. An important aspect of human experience in general is constituted by movement and motor skills. Despite the strong tradition of incorporating body-based epistemologies into the design and evaluation of new tools and techniques, the potential of interactive technologies to facilitate somatic expertise as a part of athletic training in sports has been limited. As embodied beings, movement is our primary mode of exploring and navigating the world that surrounds us. In equitation, for example, movement is a human's only way of establishing a meaningful connection with a horse. Thus, we need technologies that account for the sense modalities and coenesthetic dimensions of movement within the frame of athletic performance. The implementation of interactive technologies to mediate heightened experiential body knowledge and somatic awareness as an integral part of athletic training has not been developed yet. For example, the idea of 'imaging'—the notion that images in the mind can shape the physiology and neuromuscular behavior of the body (see Franklin, 2014) has served as a new resource for practice and experimentation in dance-related interactive systems (Raheb et al., 2019). Flow (Csikszentmihalyi, 1990) and feel (Gendlin, 1996; Podhajsky, 1991) have also been utilized as design parameters and a targeted objective in interactive movement-based systems (Isbister & Mueller, 2015). These all can be explored and integrated in interactive sports technologies to rearticulate the experiential properties of movement in athletic performance.

The exploration of movement as not solely functional, but also as an experiential modality within the domain of HCI poses significant opportunities and challenges for the production, evaluation, and utilization of interactive sports technologies. To regard the lived body so affirmatively and to investigate when and how we feel most fully aware or alive calls for a radical shift in how we frame, approach, and design interactive technologies for sports. It also inspires us to engage in research inquiry that transcends the dominant functional treatment of technologies and moves us to where there is no perceptible detachment of the user from the object of technology. This can be achieved through active engagement of affectiveness, resonances, synchronies, harmonies, and attunement—visceral states that are often experienced through movement, especially in sports. In such a shift, the primary mode of sense-making transcends the sensory apparatus, and even kinetic/kinaesthetic and affective modalities of bodily consciousness (Sheets-Johnstone, 2011), to reconcile with the very force of life itself. The practical challenge of such an endeavor lies in trying to grasp the deeper sensibilities of our senses—which do not engage with the external sensations of the visible, aural and tactile world, but with our inner selves that enable us to feel alive.

1.2.2 Mind-Body Dualism

The 'mind-body dualism' thematic tension is an extension of the quest for technological efficiency. The Cartesian tradition of treating the mind as a separate entity from the body has resulted in the mind-body dualism that has established itself as a modus operandi in Western thought. We can see this dualism reflected in the shifting approaches to HCI characterized by the three 'waves' (Bødker, 2015). The first wave of ergonomics-based approaches corresponds to the structural pairing of technology to the Body, where second wave information processing models target the cognitive capacities of the Mind. The third-wave's focus on experience serves as an attempt to reconcile these two, but still with limited practical applicability. A more integrationist approach will consider the role of technology in the mediation of mindfulness and reflection—a state in which "the mind is present in embodied everyday experience" (Varela et al., 1991, p. 22). We can see this idea represented in the notion of reflective informatics (Baumer, 2015; Núñez-Pacheco & Loke, 2015) and slow technology (Hallnäs & Redström, 2001; Odom et al., 2012). The transcendence of body-mind dualism in the field of HCI has had limited consideration within the subfield of interactive sports technologies. With some exceptions, most interactive technologies for sports limit their focus to the body. The role of reflection, imaging, felt-sense, coenesthesia, and somatic awareness as constituents of the 'body/mind' whole—and thus also athletic performance—have not been thoroughly explored yet. Interactive sports technologies as a subfield of HCI can extend itself by giving consideration to approaches that bring body and mind together. Such approaches are particularly essential to the design of supportive technologies for contact sports—where equitation serves as an example—as their practice requires a mastery of one's own deeds and actions before anything else (Rittenmeister von Oeynhausen, 1845).

In *The Primacy of Movement*, Maxine Sheets-Johnstone (2011) examines animation as the foundation of an individual's perceptual world, or a tool of perception. She argues that the sense of aliveness is grounded in movement and as such, movement is primal to perception. It is encoded in our nature to keep at focus what is moving over what is not. Her conceptualization of movement—as primal to the state of being alive—transcends the traditional mind-over-matter dualism established in Western philosophy. She places the phenomenon of movement as the bodily source of mind and cognition. Knowledge emerges from self-knowledge, and self-knowledge is created predominantly through 'moving oneself.' Richard Shusterman (2008) proposes an alternative approach that brings body and mind together through his notion of somaesthetics. It provides a unique understanding of the lived body as a "locus of sensory-aesthetic appreciation (aesthesis) and creative self-fashioning" (p.19). In contrast to classical conceptualizations of sensory perception as limited and even obstructive to intellectual perception (see Descartes, 2002), Shusterman engages the soma as "primordial instrument in grasping the world" (2008, p. 19) within which experiences arise, get shaped, and generate life-sustaining knowledge.

The idea of feel—used in philosophy, psychology, HCI, and many sports activities—is another approach that circumvents the mind-body dualism.

According to Eugene Gendlin (1993), a felt-sense is a form of inner knowing and awareness—something that forms, is distinctly there, occupies our bodies, but has a space of its own. Gendlin illustrates the notion of felt sense as experiences of diffuse feelings inside our bodies, which can be summed up as "hunches," intuition, or the idea of having a "feel" for a situation without being able to specifically verbalize the "why's" of such sensations.

The goal of much HCI in sports and athletic contexts should include developing athletic performance at least partially through heightened coenesthesia and somatic awareness. This approach acknowledges movement as a knowledge accumulating modality in recognition that movement transcends the body. Notions of embodiment that bring the body and the mind together can be mobilized to inform research knowledge and approaches toward design of technologies for sports and produce novel epistemological and computational models. From a practical perspective, the field can benefit from the diverse scholarship connecting HCI, first-person somatic perspectives, movement (Fdili Alaoui et al., 2015), and performance (Brown, 2019). Rich as the embodied approaches to HCI are, however, I have not seen them applied in a systematic way to sports and athletics—a gap not just in the theoretical contributions, but also in the everyday practice of (and in the massive industry supporting) sports and athletics. Transcending the mind-body dualism in the domain of interactive sports technologies can also contribute to the theory, everyday practice, and institutional activity of sports. The development of the notion of embodiment within interactive sports technologies can be actively linked to perspectives such as slow technology (Hallnäs & Redström, 2001; Odom et al., 2012), sensation and physical intelligence (Lobel & Bean, 2014) and physical literacy, which is "the motivation, confidence, physical competence, knowledge and understanding to maintain physical activity throughout the lifecourse ... as appropriate to each individual's endowment" (Whitehead, 2010, p. 11), through which educators can develop "embodied subjects instead of docile bodies" (Markula, 2004, p. 62). The rationale for pushing the frontiers of embodiment as a foundational approach to HCI (Dourish, 2001) by examining its application to the areas of sports and athletics is both compelling and strategic, since sports provide an important and popular avenue for addressing issues around health, wellbeing, and 'the good life.'

1.2.3 Goal-Oriented vs. Altruistic Enframings of Technology

Although this is the last thematic tension we discuss, we find it most important. Emerging from the constructive debates surrounding the ethical use of technologies in regard to animal welfare, we extend on this notion to argue that the design + application of contemporary interactive sports technologies should emanate from a deep, empathic consideration of what has been conceptualized as 'otherness' in its many forms: human diversity—with a specific focus on including individuals with disabilities in sports activities, animals, vegetation, climate, and sustainability of nature among others. In our design and use of technologies, we tend to emerge from a place of extreme goal-orientation—placing efficiency, functionality, and cost at the center of attention. This is evidenced by the significant damage we have inflicted on the environment and its ecosystems as a result of our technological progress. We have been reaching a point of no return and it is time for change. As a starting point, we can shift our anthropocentric positionality to a more altruistic, empathetic, and compassionate state of mind in an attempt to bridge the gap formed by the absence of a sense/feel for the Other. When it comes to HCI, we can create a reflective space for ethical valuation of technology design and applications through an altruistic and compassionate approach reimaging the experiential embodiment of otherness.

Chrétien (2004) argues that the first evidence of soul is the sense of touch. Touch is always relational to what is being touched and unmasks the "sheer kinetic spontaneity with which a vast and incredible diversity of animate forms comes into the world" (Sheets-Johnstone, 2011, p. 271). The immense potency of contact sports grounded predominantly in tactility seems to exemplify the kinaesthetic and affective capacities of touch which in their relational flow tap into what can be called "self/less consciousness." Buddhist interpretations of self/less consciousness offer great insight into how somatic awareness (of movements of the breath and of vital energies) can configure bodily existence (via Varela et al., 1991). It is the realization that our being, consciousness, and experiences emerge in relation to and as a part of the environment with all its constituent animate and inanimate entities. As such, our actions (also implicated in use of technologies) should arise from a stance of altruism, respect, and responsibility for the world outside ourselves. New trends in ethology, anthrozoology, and human-animal interactional studies offer ways in which we can reconsider an ethical approach to the design and use of technologies. With their fast speed of development and utilization, we believe that interactive sports technologies carry the potential of setting the trend in this direction. There are many ways forward, some of which include:

1. Integrate altruistic values within the context of HCI in order to expand empathy in technology design, utilization, application, and research.
2. Enhance the notion of *more-than-human* as a technology- and design-parameter. This is especially applicable to interactive technologies for sports which engage animals.
3. Create a reflective space for the ethical evaluation of technology design and applications.
4. Develop interactive technologies that challenge users' epistemic conditionings.
5. Apply altruism-based design approaches to the development of interactive technologies which facilitate reflection and the development of self/less consciousness.
6. Produce technologies that reduce the human imprint on the environment.

1.3 The Role of Interactive Sports Technologies in Fostering Innovation and Inclusion

In his *Phenomenology of Spirit*, Hegel proposes the genesis of a truly free self-consciousness in-and-for-itself and suggests that such free self-consciousness emerges within the essence of work whose purpose is to create rather than consume a thing (via Gadamer, 2013). As essential creators of things, technology designers have direct access to and a tremendous influence on the human condition. The utilization of technologies across all domains of life makes us consider not only functionality, but also the primary ways of understanding and examining values, meanings, and experience. The larger epistemological assumptions of design knowing and inquiry have come into existence as an extension of an aggregate body of knowledge generated within the long intellectual histories of the arts, science, and humanities. We are discovering how these traditions have shaped and continue to shape the spectrum of creative intelligence synthesized within the design process and the importance of experience and imagination within which technological artifacts are created. As creative as design practice is supposed to be, it often emerges as a product of a wide range of values, knowing, and approaches that have already been established and dominate the field. Thus, the design, experimentation, production, and theoretical underpinnings of the creative process often reproduce the biases and assumptions that come inherent within the quest for progress. It is essential that the framing and shaping of the design/research creation takes place in a critical reflective space open to interdisciplinary knowledge transfer and where epistemic understandings can be safely questioned, challenged, and transformed. The theory and practice of interactive sports technologies offer an opportunity for such safe reflective spaces where we can not only create more 'considerate' technologies, but become better beings too.

In light of these considerations, *Interactive Sports Technologies* focuses on conceptualizing and documenting current approaches, contexts, and practices of the field. It presents one of the first organized attempts at systematizing relevant knowledge and providing reflective insights that may inform current and future scholarship. By doing so, this anthology serves as a resource for better understanding the significance, benefits, and social impact of interactive sports technologies to technology design + research, and utilization. The interdisciplinary combination of perspectives, approaches, and contexts facilitates the epistemological reimagining of the field. Transcending disciplinary boundaries, this edited collection covers a diverse range of areas related to technology-physical body integration, technologies for training reactive skills, techno-methodological procedures for measuring physical performance, data collection for performance metrics, data-driven sonification as real-time feedback, augmented reality for training, augmented ball-based team games, gamified football for training and engagement, and sports technologies for parasports and social inclusion. In lieu of an afterward, we provide a syllabus for an undergraduate level course entitled 'Embodiment and Skill Acquisition in Sports Technologies.' We hope that a ready-to-use plan for such a course will inspire faculty at colleges and universities to consider, develop, propose, and—hopefully—teach such courses.

References

Antle, A. N., Marshall, P., & van den Hoven, E. (2011). Workshop on Embodied Interaction: Theory and Practice in HCI. *Proceedings of the 2011 Annual Conference Extended Abstracts on Human Factors in Computing Systems - CHI EA '11*, 5. Association for Computing Machinery. 10.1145/1979742.1979592

Baumer, E. P. S. (2015). Reflective Informatics: Conceptual Dimensions for Designing Technologies of Reflection. *Proceedings of the 33rd Annual ACM Conference on Human Factors in Computing Systems - CHI '15*, 585–594. Association for Computing Machinery. 10.1145/2702123.2702234

Bødker, S. (2015). Third-Wave HCI, 10 Years Later—Participation and Sharing. *Interactions, 22*(5), 24–31. 10.1145/2804405

Brown, C. (2019). Machine Tango: An Interactive Tango Dance Performance. *Proceedings of the Thirteenth International Conference on Tangible, Embedded, and Embodied Interaction*, 565–569. Association for Computing Machinery. 10.1145/3294109.3301263

Cairns, P., & Power, C. (2018). Measuring Experiences. In M. Filimowicz & V. Tzankova (Eds.), *New directions in third wave human-computer interaction: Volume 2—Methodologies* (pp. 61–80). Springer International Publishing. 10.1007/978-3-319-73374-6_5

Chrétien, J.-L. (2004). *The call and the response* (1st English language ed.). Fordham University Press.

Csikszentmihalyi, M. (1990). *Flow: The psychology of optimal experience* (1st ed.). Harper & Row.

Descartes, R. (2002). *Meditations on first philosophy* (S. Tweyman, Ed.). Caravan Books.

Dewey, J. (1997). *Experience and education* (1st ed.). Simon & Schuster.

Dourish, P. (2001). *Where the action is: The foundations of embodied interaction.* MIT Press.

Ellul, J. (1973). *The technological society.* Random House USA Inc.

Fdili Alaoui, S., Schiphorst, T., Cuykendall, S., Carlson, K., Studd, K., & Bradley, K. (2015). Strategies for Embodied Design: The Value and Challenges of Observing Movement. *Proceedings of the 2015 ACM SIGCHI Conference on Creativity and Cognition - C&C '15*, 121–130. Association for Computing Machinery. 10.1145/2757226.2757238

Feenberg, A. (2010). *Between reason and experience: Essays in technology and modernity.* MIT Press.

Franklin, E. N. (2014). *Dance imagery for technique and performance* (Second ed.). Human Kinetics.

Gadamer, H.-G. (2013). *Truth and method* (First paperback edition.translation revised by Joel Weinsheimer and Donald G. Marshall). Bloomsbury.

Gendlin, E. T. (1996). *Focusing-oriented psychotherapy: A manual of the experiential method.* Guilford Press.

Gendlin, E. T. (1993). Three Assertions About the Body. *The Folio, 12*(1), 21–33.

Hallnäs, L., & Redström, J. (2001). Slow Technology – Designing for Reflection. *Personal and Ubiquitous Computing, 5*(3), 201–212. 10.1007/PL00000019

Hanna, T. (1986). What Is Somatics? *Somatics Journal of the Bodily Arts and Sciences, 5*(4), 4–8.

Heidegger, M. (1996). *The question concerning technology and other essays* (W. Lovitt, Trans.). Harper and Row.

Höök, K. (2018). *Designing with the body: Somaesthetic interaction design.* The MIT Press.

Höök, K. (2010). *Transferring qualities from horseback riding to design. Proceedings of the 6th Nordic Conference on Human-Computer Interaction: Extending Boundaries.* 226–235. Association for Computing Machinery. 10.1145/1868914 .1868943

Isbister, K., & Mueller, F. "Floyd." (2015). Guidelines for the Design of Movement-Based Games and Their Relevance to HCI. *Human–Computer Interaction, 30*(3–4), 366–399. 10.1080/07370024.2014.996647

Lee, W., Lim, Y., & Shusterman, R. (2014). Practicing Somaesthetics: Exploring Its Impact on Interactive Product Design Ideation. *Proceedings of the 2014 Conference on Designing Interactive Systems - DIS '14*, 1055–1064. Association for Computing Machinery. 10.1145/2598510.2598561

Lobel, T., & Bean, J. (2014). *Sensation: The new science of physical intelligence.* Brilliance Audio; Unabridged edition.

Loke, L., & Schiphorst, T. (2018). The Somatic Turn in Human-Computer interaction. *Interactions, 25*(5), 54–5863. 10.1145/3236675

Markula, P. (2004). Embodied Movement Knowledge in Fitness and Exercise Education. In L. Bresler (Ed.), *Knowing bodies, moving minds: Towards embodied teaching and learning* (pp. 61–76). Kluwer Academic Publishers.

Marx, K., & Engels, F. (1983). *Letters on "Capital."* New Park Publications; Distributed in the U.S. by Labor Publications.

Núñez-Pacheco, C., & Loke, L. (2015). The Felt Sense Project: Towards a Methodological Framework for Designing and Crafting from the Inner Self. *Proceedings of the 21st International Symposium on Electronic Art.* ISEA, Vancouver, BC, Canada.

Nylander, S., Tholander, J., Mueller, F., & Marshall, J. (2014). HCI and Sports. *Proceedings of the Extended Abstracts of the 32nd Annual ACM Conference on Human Factors in Computing Systems - CHI EA '14*, 115–118. Association for Computing Machinery. 10.1145/2559206.2559223

Odom, W., Banks, R., Durrant, A., Kirk, D., & Pierce, J. (2012). Slow Technology: Critical Reflection and Future Directions. *Proceedings of the Designing Interactive Systems Conference on - DIS '12*, 816. Association for Computing Machinery. 10.1145/2317956.2318088

Podhajsky, A. (1991). *The complete training of horse and rider in the principles of classical horsemanship.* Wilshire Book.

Raheb, K. E., Stergiou, M., Katifori, A., & Ioannidis, Y. (2019). Dance Interactive Learning Systems: A Study on Interaction Workflow and Teaching Approaches. *ACM Computing Surveys, 52*(3), 1–37. 10.1145/3323335

Sheets-Johnstone, M. (2011). *The primacy of movement* (Expanded 2nd ed.). John Benjamins Publishing.

Shusterman, R. (2008). *Body consciousness: A philosophy of mindfulness and so-maesthetics.* Cambridge University Press.

Varela, F. J., Thompson, E., & Rosch, E. (1991). *The embodied mind: Cognitive science and human experience.* MIT Press.

Whitehead, M. (Ed.). (2010). *Physical literacy: Throughout the lifecourse* (1st ed.). Routledge.

2 Interactive Technology Integrating with the Physically Active Human Body: Learnings from Rider and Ebike Integration

Josh Andres and Florian 'Floyd' Mueller

2.1 Introduction

Interactive technology designed to support engaging in whole-body movement has mostly treated the body of the user and the interactive system as two separate entities, following teachings of traditional interaction paradigms (Dourish, 2004; Hornbæk & Oulasvirta, 2017). To further the state of technology supporting whole-body movement we, instead, propose engaging with the paradigm of human-computer integration, where the user and the system work as partners (Farooq & Grudin, 2016) and data flows between them as one whole entity (Andres et al., 2020; Mueller et al., 2020). We believe that this area deserves more investigation, especially in discovering how to offer an amplified sensation of the user's abilities as if they were "bionic". We focus on bike riding, and as such we question if we could facilitate a close human-machine experience leveraging eBikes, and explore whole-body movement to regulate the eBike's electric engine control to offer the user the sensation of increased physical strength.

Electric bikes, or "eBikes" for short, are bicycles that have an electric engine to offer the rider "power assistance" (Johnson & Rose, 2015; Jones et al., 2016). Riders enjoy the engine support as they can go faster and longer distances while engaging in physical activity. For these benefits, eBikes have become popular worldwide and across generations (Fishman & Cherry, 2016; Johnson & Rose, 2015). We note that designing interactive technology that contributes to seamlessly merging the rider's and eBike's "bodies" during rider-eBike interactions has not been explored much in sports and HCI. We propose that an integration lens can be of great importance in this context due to the rider's and the eBike's bodies working together. We believe that investigating this space can yield insights for the design of novel exertion experiences. To explore this space, which we refer to as "integrated exertion" (Andres et al., 2018), we designed "Ava, the eBike". Ava is a modified electric bike that can sense and interpret movement data continuously and in real-time from the rider's leaning forward posture to offer engine support while simultaneously playing a "turbo" sound. We focus on whole-body integration so that the rider's movements and

DOI: 10.4324/9781003205111-2

the eBike's functionality work in sync, offering the rider an amplified sensation of their abilities almost as if they were "bionic".

Our contributions are: We present the learnings of designing Ava and studying it in action as well as a set of design implications for future integrated exertion systems.

2.2 Cycling Support Technology

In this section we use two categories to draw a distinction between goal-oriented and experience-oriented cycling support technology.

2.2.1 Goal-Oriented Cycling Technology

The role of technology in cycling today is often to support bike riders to learn about their cycling performance by analysing and presenting insights from their cycling data. These insights can aid riders to improve their technique through just-in-time feedback which are often delivered via audio or a screen interface. Another approach is to offer the rider an interactive summary of the performance so that they can take a retrospective approach to the experience (Endomondo, 2020). Other works have focused on way-finding and safety, including using projections on the road to guide the rider (Dancu et al., 2015), or using a head-movement controlled helmet (Walmink et al., 2013) to indicate turning, braking and accelerating to vehicles nearby.

These goal-oriented technologies support cycling by treating the body of the rider and the eBike as separate bodies, relying on a traditional inter-action paradigm of command-response. They therefore, we believe, miss an opportunity to support whole-body integration, where the eBike becomes an extension of the rider's body as mapped in the rider's "body schema" derived from integrating visual, tactile and proprioceptive signals relating to the rider's corporeal awareness (Berlucchi & Aglioti, 1997; Maravita & Iriki, 2004).

2.2.2 Experience-Oriented Cycling Technology

Experiential aspects – in contrast to goal-oriented aspects – have also been supported through technology, such as using audio and GPS to create storytelling experiences that support wandering for the enjoyment of cycling (Rowland et al., 2009). Virtual reality and an exercise bike that can simulate cycling a virtual street while throwing newspapers to deliver to mailboxes also explored experiential aspects in a playful context (Bolton et al., 2014). Horse riding was used as inspiration to create a stationary bike that plays horse-like sounds according to the rider's cycling motion (Landin et al., 2002). And in an educational setting, exercise bikes have been connected to on-screen contents where the rider can pedal and turn to navigate the digital content, resulting in an exploratory and learning experience (Al-Hrathi et al., 2012).

Interestingly, these works offer structured and unstructured play to invite exploration (Arrasvuori et al., 2011; F. Mueller et al., 2018). The rider and system interaction in these examples begins to focus on whole-body interaction allowing the user to control and navigate digital content. This tells us that there may be a possibility to use the whole-body motion of the rider to explore experiences where the motion corresponds directly, and simultaneously, to control the eBike functionality, rather than to control virtual content. As such, we could explore an integration of the rider's whole-body motion and the eBike's functionality, resulting in a more integrated cycling experience.

2.3 Whole-Body Integration

Human-computer-integration "implies partnership [where] partners construct meaning around each other's activities, in contrast to simply taking orders" (Farooq & Grudin, 2016). In "superhuman sports" (Kunze et al., 2017), early exploration has focused on whole-body sports and mixed reality to offer users "virtual" superpowers, resulting from bodily motion, changes in the user's heart rate, and in-game score. This example provides an integration experience where the user's body is not directly actuated, nor their physical speed or strength is increased. Still, the player's surroundings and their virtual power ability are simultaneously responding to their whole-body movement. This notion of superhuman-power enabled through a whole-body integrated approach has led us to consider how we could offer the sensation of increased strength in the physical world, rather than relying on virtual simulations.

2.4 eBikes as a Research Vehicle

eBikes have gained popularity worldwide as they make cycling accessible to more people thanks to the electrical assistance (Fishman & Cherry, 2016; Plazier et al., 2017). eBikes support the benefits of engaging in physical activity and the joys of cycling socially, all while promoting environmentally conscious choices (Plazier et al., 2017). Besides their popularity worldwide, there has been limited exploration in human-computer interaction, especially on applying human-computer integration principles to imagine and explore rider-eBike integration futures.

2.5 Ava: A Case Study

Ava, the eBike, is linked to the rider's whole-body movement to allow for continuous integration between whole-body movement and engine actuation as one expression. This was inspired by previous work (Andres et al., 2016, 2018, 2019, 2020) that pointed out that in many sports users lean forward to embrace speed such as when going downhill riding bikes, skates and rollerblades. Similarly, when users need to invest more physical effort, such as when riding a bike uphill, or surfing and pushing against a curling wave, users lean forward.

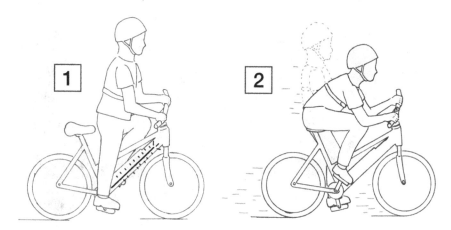

Figure 2.1 Ava senses in real time the rider's posture to 1. According to the leaning forward angle actuate the eBike's engine acceleration. 2. When the rider is likely to become wobbly, such as when resuming riding, LEDs are activated on each side of Ava's body.

This leaning forward to go faster or to invest more physical effort inspired our design, see illustration in Figure 2.1. When the user leans forward, the torso inclination angle is calculated in real-time using a body-worn gyroscope that wirelessly connects to the eBike's engine, offering an instant actuation of Ava's engine support and increases the acceleration as if the resulting extra speed came from the rider's own body leaning (Figure 2.2). For safety, we disable the engine actuation when the rider has stopped. For example, the rider might wish to reach for their water and engine actuation while doing so could be dangerous. The rider can use extra engine support by leaning intermittently, or they can remain in a leaned posture to enjoy the acceleration fully. To further the sensation of increased strength resulting in reaching faster speeds, we explored how we could engage with other senses of the rider to offer a multisensory experience that can elicit from the rider fantasy and make-believe moments that allow the user to be more engaged and enjoy the experience (Deterding, 2016). We designed a speaker placed on the eBike's handlebar. The speaker was programmed to receive the leaning data in real time and play a "turbo" sound according to the leaning intensity of the rider, aiming to contribute to the sensation of increased strength when going faster.

A second observation that inspired our design was that of bike riders standing up to pedal when resuming riding, for example, when a rider starts to cycle again after waiting at a traffic light. The effect on the rider while standing up to pedal when resuming riding is that riders can become "wobbly", or in other words, slightly unbalanced, this is especially the case with eBikes as they are often heavier than standard bikes. This observation led us to use movement data from the gyroscope worn around the rider's chest to infer, in real time,

Figure 2.2 Ava's system with selected components: a) The whole bike with added components; b) 3D printed container housing Raspberry Pi Model B to process leaning data and signal to the engine controller; c) Mounted wireless Bluetooth speaker; d) Button for sound selection.

when the rider is standing up in order to turn on two LED strips on the sides of the eBike's body. The LEDs light up to indicate to nearby vehicles that the rider is likely to be wobbly, making them more aware of the presence of the rider and potentially contributing to the safety of the rider (Figure 2.2).

2.6 The Experience of Integrating the Rider's Body with the System as One Continuous Cycling Expression

We conducted a study around Ava. Twenty two participants (F = 10, M = 12), aged between 24 and 55 (M = 36.4, SD = 9.4) from a medium size city in the

Asia Pacific hosted Ava for two weeks each and they could cycle during this time as much or as little as they wished. At the start of the two weeks, participants received a tutorial on calibrating and placing the gyroscope correctly on their chest. Participants reported after the two weeks that they exerted themselves while cycling, using their whole body and experimenting with leaning forward to embrace the speed, resulting in igniting thrill and exciting moments. They praised Ava as it provided participants with an intuitive way to operate the engine support in contrast to standard eBikes that use a variety of control mechanisms such as a throttle. Participants referred to using whole-body motion to actuate engine support as a way in which they did not need to "think" about how to control the system because it felt as if they were simply moving their body. This is particularly interesting for us because besides the integration with the user's body it allows the rider to keep their hands on the handlebars, in contrast to having to change the position of their hands to operate a button or look at a screen – which can distract from the experience of cycling and break the sensation of being one with the system.

Most participants commented on the turbo sound and described it as affording a "superpower" sensation when leaning forward, because they felt as if the power came from within their body. They described leaning and the turbo sound as if they were "motor racing". They commented that even when they were tired, they could still invoke the superpower as a boost, especially on challenging road segments or segments that lend themselves to embrace speed. There are important considerations for integrative sports technology in this comment, as the human body has performance limits but when paired so intimately with technology as one, it appears possible that the user can continue to push their physical limits by drawing from this digital extension of their body, fuelled by moments of make-believe where they exaggeratedly take a curve while embracing the extra power.

The participants also reported on occasions where Ava responded as if it was possessed. This situation occurred because in some cases participants did not calibrate the gyroscope from their upright starting position, resulting in either having to lean very low to trigger the acceleration or triggering the acceleration accidentally leaning slightly. This was particularly "scary" while going downhill and not wanting to increase extra power. In other cases, participants chose very steep roads to try Ava, and as they leaned forward to pedal the uphill road, they triggered the acceleration which was supporting in the context of the experience, but it was not the intention of the rider. These moments highlight a couple of considerations for technology integration in sports. First, when the system acts in a way that is not expected by the user, it breaks the sensation of integration because it draws their attention to the system as a separate entity. Importantly, this separation between the bodies is amplified when the system acts against the intent of the user, such as in the example when the user wanted to go downhill slowly, but Ava kept accelerating.

Second, even though the user in the uphill example did not intend to trigger the acceleration, the system did so, and it helped them to climb up.

Integrative sports technology systems could derive context of the experience, for example through other data sources such as road inclination in this case, to assist the user as a partner with the trade-off that the system breaks from being one with the user and transitions to being a supporting teammate.

We can envision that, in the future, using data sources from which we can infer a user's state, such as a heart rate, could indicate user fatigue, and this fatigue could programmatically mean more or less contribution from the system as a partner in a sports context. Users of integrative sports systems could define such bodily fatigue thresholds so that the integration system acts on the fatigue and momentarily leads tasks within the experience while giving a break to the user to recover from fatigue.

2.7 Integration Learnings

We have grouped aspects that we believe contributed to making our integration successful and here we share the insights as practical opportunities that other researchers and designers can leverage in future sport integrations.

2.8 Movement

In this section we describe movement qualities of the user in relation to system design choices to support integration.

1. *Self-attribution from existing movements to pair human-machine bodies*

We used an existing movement used by riders already to embrace speed in the context of cycling. The movement was leaning forward, and in real-time as the rider leaned, we actuated the engine gradually. This afforded a sensation, that the extra acceleration was being activated as the rider was performing the movement without delay, resulted in riders self-attributing the acceleration as if it came from their body to go faster.

We believe that real-time system actuation to facilitate a user's self-attributing system output as if it came from their body can play a key role because it blurs the gap between the human-input and the machine-output. This real-time actuation can afford the user to experience their movement intent seamlessly as one with the machine.

2. *Map movement to machine actuation to facilitate integration*

We used the rider's leaning posture when cycling bikes that have a flat handlebar (Figure 2.2). The reason for choosing this bike design is that (1) it is a popular design that a lot of people are familiar with, (2) choosing one design allowed us to explore the rider-ebike interactions more deeply based on how various users interact with the same system, and (3) we could design our system focusing on one bike style and the leaning posture angle that riders often take as other bike designs such as a drop-handle bike invite lower posture angles from the user.

The intention behind focusing on one bike style and using the specific leaning posture angle as a design resource was to design for an experience that promotes rider-bike integration. We believe that if the system would have actuated the engine independently of the user's movement, it could break the illusion of being integrated with the rider's body. A "body schema" refers to the corporeal awareness or "map" of our body shape and posture (Maravita & Iriki, 2004), this map is constantly updated in our brain by successively integrating visual, tactile and proprioceptive signals relating to our corporeal awareness. Objects with a systematic relation to the body can be absorbed into the body schema as extensions of our body. Therefore, the eBike can become an extension of the rider's body as they cycle (Berlucchi & Aglioti, 1997; Maravita & Iriki, 2004). Importantly, any such inclusions into the body schema are often temporary; once the rider is no longer on the eBike, they do not experience the eBike as an extension of their body, nor do they experience it as a lost limb.

In our case study we noticed that the cycling skill can be an integration factor when experiencing the system. More experienced cyclists appeared to be able to transfer their foundational movement and balance skills to the bike faster than less experienced riders. This may be because less experienced riders seemed to more laboriously try to adjust to the system, resulting in more "thinking" about how to use the bike rather than spontaneously moving as one with the bike. Recent work in human-computer integration expands on this observation relating to the user's skills and integration with the system by referring to conscious and subconscious attention (Spiegel, 2019). In simpler terms, as a user becomes more skilled, their actions with the system can become almost automatic or "subconscious", rather than being the focus of attention or "conscious" process where the user is thinking about how to perform an action.

3. *Design the actuation support to respond according to the movement speed to facilitate moving simultaneously with the machine as one continuous expression*

During the study, we experimented with various acceleration responses: from a strong acceleration response when leaning forward to a weak engine support response. Calibrating how much engine support to provide resulted in different experiences. For example, if the engine support was too weak, it resulted in the user wondering if the system was discharged, and it also caused users to remain leaned to get the speed they wished to reach. On the other hand, when the engine support responded by providing a strong actuation, it turned the experience into a choppy, difficult to control, and disconnected from the rider's lean forward movement experience. As such, finding the sweet spot of engine support response to whole-body movement can play a key role in maintaining the illusion of integration.

2.9 Amplification

In this section we describe how we used various sensorial qualities to amplify the sensation of being integrated with the system.

4.   *Exploit visceral qualities to ignite imaginaries of integration*

It takes time for users to reach a state of integration. To prepare for and assist with the process of integration, we borrowed from the model of three levels of design (Norman, 2004). Specifically, we borrowed from "visceral" design, the first level of the model, that refers to the perceptible qualities of the system and the associations it can evoke from a user before the user tries out the system. Visceral design can serve as a potent precursor of the experience that influences and creates expectations of what the system can do, igniting imaginaries of being integrated with the system. Following this insight, we offered users a sporty and futuristic eBike to build on the visceral aspects of the experience, considering the form factor, the colours of the eBike, and the placement of the LED lights to create a futuristic look. In the design space of integration and sports, we believe that visceral design is an under-explored resource that future integration designs should leverage to support amplifying the sensation of being integrated with the system.

5.   *Amplify the sensation of being integrated with the system by engaging other senses that complement the experience*

We used different sounds that the rider could choose from; the selected sound was played simultaneously as the rider leaned forward. The sound allowed riders to experience an amplification of the sensation of acceleration as they felt that they were going faster. Participants particularly mentioned the "turbo" sound as appreciating how it complemented the experience of accelerating.

We were initially unsure about using sound as we wanted to keep the experience of cycling as pure as possible without any distractions from cycling. The key insight from a preliminary experiment was that, similarly to actuation support, calibration was needed to match the leaning motion in real-time to become a complementary sensorial layer in the experience. Consider, for example, a car: when pressing the acceleration pedal there are accompanying sensorial qualities, from the sound of the engine revving proportionally to how far and how long the pedal is pushed for to the sensation of going faster. Failing to match the leaning forward motion in real-time caused the sound to be experienced as an external add-on that makes the sound feel "fake" to the experience. We conclude that the turbo sound working simultaneously with the engine support actuation was essential to promote make-believe moments, such as, "the extra speed comes from my body", and "it makes me feel like I'm on a turbo eBike and I appreciate going fast".

We believe that engaging additional sensory input in a complementary manner to the experience can promote make-believe moments, such as being stronger and faster.

2.10 Looking Forward: Integrating Our Bodies with Technology in a Sports Context

Our work shows that supporting whole-body exertion through a human-computer integration approach to extend the user's abilities is possible. The resulting novel experiences can afford users "superpowers" to feel "stronger" when designed carefully. We see real-time sensing and processing data for whole-body integration systems as a glue that binds the user's body and their sensorial qualities with computing machinery functionality to enable integration.

We believe that there is an opportunity for more design of integrative technology to extend the user's abilities in a sporting context. For example, focusing on levelling up the playing field between more able and less able bodies, more experienced and less experienced players, and younger and older bodies, potentially affording physical activity experiences to a broader audience. This future is promising to us, as it can foster physically and socially healthy interactions that nurture positive values.

We believe that in the near future whole-body integration experiences could provide safe, entertaining and novel ways to interact by drawing from other bodily data beyond movement. For example, using neurological data corresponding to peripheral vision to assist riders to respond to situations faster by regulating engine support (Andres et al., 2020), using traffic light data to assist riders in catching traffics lights on green by regulating the speed (Andres et al., 2019), and using GPS data to regulate engine support ahead of hills to offer a challenging or relaxing cycling experience (De La Iglesia et al., 2018). eBikes are fascinating vehicles to explore whole-body integration experiences due to the electrical engine that can be programmed to respond to different data in real-time while the user exerts.

The learnings from these works have not yet been evaluated in other personal systems that also afford exertion opportunities and have an electric engine, such as eSkates, Segways, or exoskeletons. We believe that further investigation of the takeaways presented using these exertion-inviting-electric-engine vehicles could provide insights to further our understanding on how to design for whole-body integration experiences.

2.11 Conclusion

This chapter has offered a perspective on bodily integration sports technology using an eBike that utilised movement data to integrate with the rider's body. Our reflections on this work highlight practical opportunities to design for and support bodily integration. This chapter points out how designers and researchers of interactive technology can design whole-body integration experiences where the user is "one" with the system and their abilities are extended. We believe that integration sports technology can offer a future not only to design for extended abilities but to design technology complementary to our bodies' ability to promote a physically and socially healthy future that nurtures positive values.

References

Al-Hrathi, R., Karime, A., Al-Osman, H., & El Saddik, A. (2012). Exerlearn bike: An exergaming system for children's educational and physical well-being. *2012 IEEE International Conference on Multimedia and Expo Workshops*, 489–494. IEEE.

Andres, J., de Hoog, J., & Mueller, F. "Floyd." (2018). "I had super-powers when ebike riding" Towards understanding the design of integrated exertion. *Proceedings of the 2018 Annual Symposium on Computer-Human Interaction in Play (CHI PLAY '18)*, 19–31. New York, NY, USA: Association for Computing Machinery. 10.1145/3242671.3242688

Andres, J., De Hoog, J., Von Känel, J., Berk, J., Le, B., Wang, X., Brazil, M., & Mueller, F. (2016). Exploring human: Ebike interaction to support rider autonomy. *Proceedings of the 2016 Annual Symposium on Computer-Human Interaction in Play Companion Extended Abstracts (CHI PLAY Companion '16)*, 85–92. New York, NY, USA: Association for Computing Machinery. 10.1145/2 968120.2987719

Andres, J., Kari, T., Von Kaenel, J., & Mueller, F. (2019). Co-riding with My eBike to get green lights. *Proceedings of the 2019 on Designing Interactive Systems Conference(DIS '19)*, 1251–1263. New York, NY, USA: Association for Computing Machinery. 10.1145/3322276.3322307

Andres, J., Schraefel, M. C., Semertzidis, N., Dwivedi, B., Kulwe, Y. C., von Kaenel, J., & Mueller, F. F. (2020). Introducing peripheral awareness as a neurological state for human-computer integration. *Proceedings of the 2020 CHI Conference on Human Factors in Computing Systems*, 1–13. New York, NY, USA: Association for Computing Machinery. 10.1145/3313831.3376128

Arrasvuori, J., Boberg, M., Holopainen, J., Korhonen, H., Lucero, A., & Montola, M. (2011). Applying the PLEX framework in designing for playfulness. *Proceedings of the 2011 Conference on Designing Pleasurable Products and Interfaces*, 1–8.

Berlucchi, G., & Aglioti, S. (1997). The body in the brain: Neural bases of corporeal awareness. *Trends in Neurosciences*, 20(12), 560–564.

Bolton, J., Lambert, M., Lirette, D., & Unsworth, B. (2014). PaperDude: A virtual reality cycling exergame. In *CHI'14 Extended Abstracts on Human Factors in Computing Systems*, 475–478. New York, NY, USA: Association for Computing Machinery. 10.1145/2559206.2574827

Dancu, A., Vechev, V., Ünlüer, A. A., Nilson, S., Nygren, O., Eliasson, S., Barjonet, J.-E., Marshall, J., & Fjeld, M. (2015). Gesture bike: Examining projection surfaces and turn signal systems for urban cycling. *Proceedings of the 2015 International Conference on Interactive Tabletops & Surfaces (ITS '15)*, 151–159. New York, NY, USA: Association for Computing Machinery. 10.1145/2817721.2817748

De La Iglesia, D., De Paz, J., Villarrubia González, Gabriel, Barriuso, Alberto, Bajo, Javier, Corchado, Juan. (2018). Increasing the intensity over time of an electric-assist bike based on the user and route: The bike becomes the gym. *Sensors, 18*(1), 220. https://pubmed.ncbi.nlm.nih.gov/29342900/

Deterding, S. (2016). Make-believe in gameful and playful design. In *Digital make-believe* (pp. 101–124). Springer, Cham.

Dourish, P. (2004). What we talk about when we talk about context. *Personal Ubiquitous Computing, 8*(1), 19–30. 10.1007/s00779-003-0253-8

Endomondo | Free your endorphins running, walking, cycling and more. (2020). Retrieved December 31, 2020, from https://www.endomondo.com/

Farooq, U., & Grudin, J. (2016). Human-computer integration. *Interactions, 23*(6), 26–32. 10.1145/3001896

Fishman, E., & Cherry, C. (2016). E-bikes in the mainstream: Reviewing a decade of research. *Transport Reviews, 36*(1), 72–91.

Hornbæk, K., & Oulasvirta, A. (2017). What is interaction? *Proceedings of the 2017 CHI Conference on Human Factors in Computing Systems (CHI '17)*, 5040–5052. New York, NY, USA: Association for Computing Machinery. 10.1145/3025453.3025765

Johnson, M., & Rose, G. (2015). *Safety implications of e-bikes*. Royal Automobile Club of Victoria (RACV). https://www.racv.com.au/membership/member-benefits/expert-advice/advocacy-for-members/research-and-reports.html

Jones, T., Harms, L., & Heinen, E. (2016). Motives, perceptions and experiences of electric bicycle owners and implications for health, wellbeing and mobility. *Journal of Transport Geography, 53*, 41–49.

Kunze, K., Minamizawa, K., Lukosch, S., Inami, M., & Rekimoto, J. (2017). Superhuman sports: Applying human augmentation to physical exercise. *IEEE Pervasive Computing, 16*(2), 14–17.

Landin, H., Lundgren, S., & Prison, J. (2002). The iron horse: A sound ride. *Proceedings of the Second Nordic Conference on Human-Computer Interaction (NordiCHI '02)*, 303–306. New York, NY, USA: Association for Computing Machinery. 10.1145/572020.572075

Maravita, A., & Iriki, A. (2004). Tools for the body (schema). *Trends in Cognitive Sciences, 8*(2), 79–86.

Mueller, F., Byrne, R., Andres, J., & Patibanda, R. (2018). Experiencing the body as play. *Proceedings of the 2018 CHI Conference on Human Factors in Computing Systems*, Paper 210, 1–13. New York, NY, USA: Association for Computing Machinery. 10.1145/3173574.3173784

Mueller, F. F., Lopes, P., Strohmeier, P., Ju, W., Seim, C., Weigel, M., Nanayakkara, S., Obrist, M., Li, Z., Delfa, J., Nishida, J., Gerber, E. M., Svanaes, D., Grudin, J., Greuter, S., Kunze, K., Erickson, T., Greenspan, S., Inami, M., … Maes, P. (2020). Next Steps for Human-Computer Integration. *Proceedings of the 2020 CHI Conference on Human Factors in Computing Systems*, 1–15. New York, NY, USA: Association for Computing Machinery. 10.1145/3313831.3376242

Norman, D. A. (2004). *Emotional design: Why we love (or hate) everyday things*. Basic Civitas Books.

Norman, D. A. (2009). *The design of future things. Basic books*.

Plazier, P. A., Weitkamp, G., & van den Berg, A. E. (2017). "Cycling was never so easy!" An analysis of e-bike commuters' motives, travel behaviour and experiences using GPS-tracking and interviews. *Journal of Transport Geography, 65*, 25–34.

Rowland, D., Flintham, M., Oppermann, L., Marshall, J., Chamberlain, A., Koleva, B., Benford, S., & Perez, C. (2009). Ubikequitous computing: Designing interactive experiences for cyclists. *Proceedings of the 11th International Conference on Human-Computer Interaction with Mobile Devices and Services (MobileHCI '09)*, Article 21, 1–11. New York, NY, USA: Association for Computing Machinery. 10.1145/161385 8.1613886

Spiegel, B. (2019). *The upper half of the motorcycle: On the unity of rider and machine.* Motorbooks.

Walmink, W., Chatham, A., & Mueller, F. (2013). Lumahelm: An interactive helmet. In *CHI'13 Extended Abstracts on Human Factors in Computing Systems*, 2847–2848New York, NY, USA: Association for Computing Machinery. 10.1145/2468356.2479542

3 Technologies and Methodological Procedures for Measuring Physical Performance in a Velocity-Controlled Resistance Training Setting

Diogo L. Marques, Henrique P. Neiva,
Daniel A. Marinho, and Mário C. Marques

3.1 Introduction

The development of new sports technologies is continuously expanding worldwide to enhance and support the athlete's and the team's performance. The requirement of a high-performance level in every sporting event has led coaches to implement new technological equipment during their working routine to control the athlete's training load (Marinho & Neiva, 2020). For example, in recent years, new technologies in resistance training settings have been challenging coaches to switch the traditional prescription based on fixed volumes and intensities for a more refined method through monitoring movement velocity (González-Badillo et al., 2011; González-Badillo & Sánchez-Medina, 2010). Using the movement velocity as the primary variable to assess, prescribe, and monitor the athletes' training load is termed velocity-controlled resistance training (VCRT) (González-Badillo et al., 2011; González-Badillo, Sánchez-Medina, et al., 2017; González-Badillo & Sánchez-Medina, 2010). Due to the load-velocity relationship, coaches can estimate the one-repetition maximum (1RM) from the values of movement velocity, make inferences in real-time about the athlete's level of effort, and monitor fatigue during sessions (González-Badillo et al., 2011; González-Badillo & Sánchez-Medina, 2010).

As a result of the positive effects on the physical performance of athletes (Pareja-Blanco et al., 2017; Rodríguez-Rosell et al., 2020), VCRT has gained popularity worldwide, and the number of velocity measurement devices has significantly increased in the last few years (Tomasevicz et al., 2020). Nowadays, coaches and researchers have at their disposal a wide range of technologies for measuring lifting velocity, including linear velocity/position transducers, three-dimensional (3D) motion analysis systems, infrared optoelectronic cameras, inertial measurement units, or smartphone applications (Courel-Ibáñez et al., 2019; Lorenzetti et al., 2017; Martínez-Cava et al., 2020; Pérez-Castilla et al., 2019; Pérez-Castilla et al., 2019; Thompson et al., 2020). According to the literature, linear velocity/position transducers, 3D motion analysis systems, and infrared optoelectronic cameras seem to be the most accurate and recommended technologies for measuring lifting velocity during

DOI: 10.4324/9781003205111-3

resistance exercises (Courel-Ibáñez et al., 2019; Lorenzetti et al., 2017; Martínez-Cava et al., 2020; Pérez-Castilla et al., 2019; Pérez-Castilla et al., 2019; Thompson et al., 2020). Conversely, inertial measurement units seem to present the lowest accuracy (Courel-Ibáñez et al., 2019; Lorenzetti et al., 2017; Pérez-Castilla et al., 2019; Pérez-Castilla et al., 2019; Thompson et al., 2020), while the findings regarding smartphone applications' reliability are controversial. Although some research recommends smartphone applications to track the lifting velocity in VCRT settings (Pérez-Castilla et al., 2019; Pérez-Castilla et al., 2019; Thompson et al., 2020), others do not support its use (Courel-Ibáñez et al., 2019; Kasovic et al., 2020; Martínez-Cava et al., 2020; Sánchez-Pay et al., 2019). Nevertheless, the rationale for using one device instead of the other should be based on the available financial resources, type of context (training *vs.* laboratory settings), and the degree of accuracy and sensitivity to measure and identify significant changes in athletic performance (Hernández-Belmonte et al., 2020; Pérez-Castilla et al., 2019; Pérez-Castilla et al., 2019).

Besides the device's choice to use in a VCRT setting, coaches must adopt validated velocity-controlled testing protocols and appropriate statistical procedures to analyze the data. The most used evaluation protocol is the progressive loading test, which consists of a gradual load increment while monitoring the lifting velocity until the athlete achieves the 1RM (Sánchez-Medina et al., 2010). Although several studies have used this test to establish the load-velocity relationship in the bench press (González-Badillo & Sánchez-Medina, 2010), back squat (Sánchez-Medina et al., 2017), or deadlift (Benavides-Ubric et al., 2020), to date, no study summarized the key-points of these testing protocols. Therefore, gathering these data is essential to define guidelines for strength and conditioning coaches who aim to conduct a progressive loading test using these exercises. Also, to give a broader perspective of the potential applications of VCRT, it is critical to discuss the velocity variables collected during the lifts and the statistics used to model the load-velocity relationship.

Therefore, in this chapter, we aimed to provide an overview of the different technologies and methodological procedures used to measure the athlete's performance in a VCRT setting. A first reference is given to different technologies for measuring lifting velocity, highlighting their respective validity and reliability. We then summarize velocity-controlled testing protocols on classic resistance exercises, the mechanical variables that can be measured, and the statistics used to model the load-velocity data. Finally, we provide a practical example describing how to implement VCRT following specific steps using a reference device for measuring lifting velocity and its respective software.

3.2 Technologies for Measuring Movement Velocity

The most used technologies for measuring movement velocity in VCRT settings are linear velocity transducers (González-Badillo & Sánchez-Medina, 2010), linear position transducers (Muniz-Pardos et al., 2019), 3D motion analysis systems (Martínez-Cava et al., 2020), infrared optoelectronic cameras (García-Ramos, Pérez-Castilla, et al., 2018), inertial measurement units (Orange et al., 2019), and smartphone applications (Peart et al., 2019). Linear velocity transducers directly measure the velocity by recording electrical signals proportional to cable velocity. On the other hand, linear position transducers, 3D motion analysis systems, and infrared optoelectronic cameras indirectly measure the velocity by numerical differentiation of displacement-time data. Inertial measurement units indirectly measure the velocity by numerical integration of the acceleration-time data. Regarding smartphone applications, some use the incorporated video camera to indirectly measure the velocity by dividing the lift's vertical distance by the concentric action time, and others use different embedded sensors, like chronometers, to indirectly measure the movement velocity using specific equations. Several devices for measuring movement velocity using each of the technologies will be presented in the following section.

3.2.1 Movement Velocity Measurement Devices

3.2.1.1 Linear Velocity and Position Transducers

Linear velocity and position transducers are electromechanical measuring systems equipped with a tethered cable attached to the athlete or the equipment (e.g., machines or free-weights), which extends and collects the data vertically (González-Badillo et al., 2017).

One device that incorporates a linear velocity transducer is the T-Force Dynamic Measurement System (Ergotech Consulting, Murcia, Spain). It consists of a transducer interfaced to a computer utilizing a 14-bit resolution analog-to-digital data acquisition board and custom software (González-Badillo & Sánchez-Medina, 2010). The device records the velocity at a sampling frequency of 1000 Hz, and subsequently, it is smoothed with a fourth-order low-pass Butterworth filter with a cutoff frequency of 10 Hz. The device's validity and reliability were established by comparing it with a high-precision digital height gauge (Mitutoyo HDS-H60C; Mitutoyo, Corp., Kawasaki, Japan) (Sánchez-Medina & González-Badillo, 2011). After comparing 18 T-Force devices, the data revealed a mean relative error in the velocity measurements of $< 0.25\%$ and an absolute error of the displacement of $< \pm 0.5$ mm. Additionally, when concomitantly performing 30 repetitions using 2 T-Force devices (range of velocities: 0.3–2.3 m·s^{-1}), the inter-device relative reliability (i.e., intra-class correlation coefficient) varied between 0.99 and 1.00, while the absolute reliability (i.e., coefficient of variation) varied between 0.57% and 1.75% (Sánchez-Medina & González-Badillo, 2011). These results suggested an excellent level of reliability and consistency of the T-Force for measuring lifting velocity during linear displacements. Given its high validity and reliability, several studies have used the T-Force as the gold standard or reference instrument to validate other devices embedded with different technologies (Courel-Ibáñez et al., 2019; García-Ramos et al., 2018; Garnacho-Castaño et al., 2015; Martínez-Cava et al., 2020). Therefore, from a sports performance and scientific perspective, the T-Force stands out as a highly valid, reliable, and sensitive device for measuring lifting velocity during resistance exercises.

Regarding linear position transducers, there is a multitude of devices using this technology, such as the GymAware Power Tool (Kinetic Performance Technology, Canberra, Australia) (Banyard et al., 2017), Tendo Weightlifting Analyzer System (TENDO Sports Machines; Trencin, Slovak Republic) (Garnacho-Castaño et al., 2015), Chronojump Linear Encoder (Chronojump Boscosystem, Barcelona, Spain) (Pérez-Castilla et al., 2019), Musclelab Linear Encoder (Ergotest Technology AS, Porsgrunn, Norway) (Bosquet et al., 2010), SmartCoach Power Encoder (SmartCoach Europe, Stockholm, Sweden) (Balsalobre-Fernández et al., 2017), or Vitruve (formerly Speed4Lift) (Vitruve Linear Encoder, Madrid, Spain) (Pérez-Castilla et al., 2019). As with the T-Force system, all devices use a tethered cable attached to the athlete or the equipment to record the lifting velocity. However, some devices transmit the data via wireless to a tablet, while others to a computer through an interface. For example, the GymAware, which time-stamps the displacement data at 20 ms time points and down-samples it to 50 Hz, transfers the data via Bluetooth to a tablet (Banyard et al., 2017). The GymAware seems to be valid and reliable to measure a wide range of velocities during the free-weight back squat and bench press and the deadlift (Banyard et al., 2017; Chéry & Ruf, 2019; Grgic et al., 2020; Orange et al., 2020). However, the reliability of the GymAware to measure the lifting velocity in the free-weight back squat and bench press might decrease at loads < 40% 1RM (Orange et al., 2020), while for the deadlift, the reliability is questionable at loads ≥ 90% 1RM (Chéry & Ruf, 2019; Grgic et al., 2020). Consequently, coaches and researchers should be aware of potentially misleading velocity results at these loads.

A lifting measurement device that transfers the data through an interface to a computer is the Chronojump. This device calculates the velocity through the numerical differentiation of time-displacement data at a sampling frequency of 1000 Hz. Based on a validation study, the Chronojump presented moderate to good reliability results for measuring lifting velocity at loads between 45% and 85% 1RM (Pérez-Castilla et al., 2019). However, the device's reliability to measure velocities at loads < 45% 1RM and > 85% 1RM is unknown. Therefore, coaches and researchers should be cautious when interpreting the velocity data at this spectrum of relative loads.

3.2.1.2 3D Motion Analysis Systems

Some sports sciences literature considers the 3D motion analysis system as the gold standard or reference instrument for measuring lifting velocity (Goldsmith et al., 2019; Lorenzetti et al., 2017; Pérez-Castilla et al., 2019;

Tomasevicz et al., 2020). Examples of 3D motion systems used in VCRT literature include the Vicon System (Vicon Motion Systems, Oxford, UK) (Lorenzetti et al., 2017), STT (STT system, Basque Country, Spain) (Martínez-Cava et al., 2020), Optotrak Certus Motion Capture System (Northern Digital Inc., Ontario, Canada) (Goldsmith et al., 2019), Trio-OptiTrack System (V120:Trio, OptiTrack, NaturalPoint, Inc.) (Pérez-Castilla et al., 2019), Qualisys Motion Capture System (MCU 240, Gothenburg, Sweden) (Tomasevicz et al., 2020), and Elite Form Tracking System (EliteForm, Lincoln, Nebraska, USA) (Mosey et al., 2018; Tomasevicz et al., 2020). Although most 3D motion systems analyze full-body movements (e.g., gait and running analysis), the Elite Form was specifically designed to track the movement velocity and be integrated into a VCRT setting. This system consists of two 3D cameras with a sampling frequency of 30 Hz positioned on the top of a squat rack interfaced with a touch-screen computer and respective software. It is a non-invasive system because it does not use cables attached to the barbell or the athlete and provides the movement velocity data in real-time. Although the Elite Form seems reliable to measure velocities at loads ≥ 65% 1RM in different exercises (e.g., power clean, deadlift, bench press, back/front squat), its accuracy with low loads and jumping exercises are questionable (Mosey et al., 2018; Tomasevicz et al., 2020). As Mosey et al. (2018) suggested, the measurement error in lifting velocity for low loads and fast velocities might be caused by high movement variability rather than a system-specific inaccuracy. In fact, when using low-to-moderate loads, there is a final phase of the lift where the athletes start to decelerate by exerting an opposite force to the movement termed the braking phase (Sánchez-Medina et al., 2010). Consequently, for loads ≤ 70% 1RM, it is recommended to analyze the mean velocity values that only consider the propulsive phase (i.e., when the barbell acceleration > acceleration due to gravity), rather than the mean velocity values that also include the braking phase (Sánchez-Medina et al., 2010). Therefore, coaches should be encouraged to analyze the lifting velocity's propulsive phase using low-to-moderate loads and high velocities.

3.2.1.3 Infrared Optoelectronic Cameras

A lifting velocity measurement device that uses this type of technology is the Velowin (Velowin, Deportec, Murcia, Spain). It consists of a 2D single-infrared camera system interfaced to a computer, which directly tracks the displacement of a reflective marker attached to a barbell at a sampling frequency of 500 Hz. The software calculates the lifting velocity by numerical differentiation of time-displacement data. According to several validation studies, the Velowin seems to be valid and reliable to measure the lifting velocity during the free-weight back squat (20–70 kg; < 30%–90% 1RM), Smith machine half squat, and bench press (40%, 60% and 80% 1RM), and loaded countermovement jump (3.5–43.5 kg) (García-Ramos, Pérez-Castilla, et al., 2018; Laza-Cagigas et al., 2019; Muniz-Pardos et al., 2019; Peña García-Orea et al., 2018). Nevertheless, coaches should be cautious when interpreting the movement velocity results when using different exercises, including those performed on resistance machines (e.g., leg press, chest press) and relative loads > 90% 1RM. Besides its accuracy for measuring the lifting velocity at specific loads in the mentioned exercises, the Velowin does not use cables attached to the equipment as other technologies, thus facilitating its use and eliminating the risk of cable rupture (Laza-Cagigas et al., 2019). Therefore, this device seems to be a valid and reliable option for coaches to implement in a VCRT setting.

3.2.1.4 Inertial Measurement Units

Several movement velocity measurement devices incorporate this technology, including the PUSH Band (PUSH band, PUSH, Inc.) or the Beast Sensor (Beast sensor, Beast Technologies Srl., Brescia, Italy). For example, the PUSH band consists of a wearable (armband) wireless device embedded with a 3-axis accelerometer and a gyroscope. It calculates the velocity through the numerical integration of time-acceleration data at a sampling frequency of 200 Hz. Although some studies suggest the PUSH Band as a valid and reliable device for measuring lifting velocity in the Smith machine full-back squat (20–70 kg) (Balsalobre-Fernández et al., 2016) or dumbbell bicep curl and shoulder press (4.54 and 6.82 kg) (Sato et al., 2015), others studies questioned its validity and reliability (Banyard et al., 2017; Courel-Ibáñez et al., 2019; Orange et al., 2019; Pérez-Castilla et al., 2019; Pérez-Castilla et al., 2019). Specifically, both in the free-weight or Smith machine full-back squat and bench press, several authors reported high measurement errors of lifting velocity across a broad spectrum of relative loads (Banyard et al., 2017; Courel-Ibáñez et al., 2019; Pérez-Castilla et al., 2019). The low validity and reliability were also corroborated by the Beast Sensor device (Pérez-Castilla et al., 2019; Thompson et al., 2020). Consequently, it might be suggested that the inertial measurement units' current configurations for measuring lifting velocity must be used carefully. Although these devices present an important advance in sports performance evaluation, researchers and companies should continuously improve their measurement systems to provide accurate, sensitive, and consistent results, especially in VCRT settings.

3.2.1.5 Smartphone Applications

Over the last few years, there has been a growing increase in smartphone applications to collect kinetic and kinematic data either using the embedded video camera or other sensors such as the accelerometer or global positioning system (Peart et al., 2019). In VCRT settings, coaches can use several applications for measuring lifting velocities, such as My Lift (formerly Powerlift) (Balsalobre-Fernández et al., 2018), iLoad (de Sá et al., 2019), or

Iron Path (Kasovic et al., 2020). My Lift and Iron Path applications use the video camera to record the lifts. The iLoad uses the smartphone's chronometer to record the time needed to perform the lifts, and then the application automatically calculates the velocity. When using My Lift application, after the lifting is recorded, the user selects two frames: the beginning and end of the concentric phase. After identifying the frames, the application calculates the lifting velocity by dividing the distance by the time (from frames 1 to 2). Several studies analyzed My Lift application's validity and reliability to measure the lifting velocity during resistance exercises at a wide range of loads, and the findings are contradictory. While some researchers found small measurement errors (Balsalobre-Fernández et al., 2017, 2018; Pérez-Castilla et al., 2019), others observed large measurement errors, especially at low loads and high velocities (Courel-Ibáñez et al., 2019; Martínez-Cava et al., 2020; Sánchez-Pay et al., 2019). Consequently, from a scientific and elite sports perspective, researchers and coaches should be aware that the application might not meet the desired standard of accuracy and sensitivity to effectively determine strength changes and monitor physical performance (Martínez-Cava et al., 2020). Nevertheless, the application can be affordable for recreational athletes due to its low cost and ease of use.

3.3 Methodological Procedures

This section outlines some velocity-controlled testing protocols using the bench press, back squat, and deadlift. For each exercise, we present a summary of testing procedures based on the available VCRT literature. We then mention the velocity variables that can be collected and the usual statistics to model load-velocity equations.

3.3.1 Velocity-Controlled Testing Protocols

To conduct a velocity-controlled test protocol, coaches adopt a progressive or incremental loading test. It consists of a gradual load increment while monitoring the lifting velocity until the athlete achieves the 1RM (Sánchez-Medina et al., 2010). In the bench press and back squat, the test is usually conducted in a Multipower or Smith machine, which allows the device cable to displace in a fully vertical direction. Although it can also be performed using freeweights, the data collected by the lifting velocity measurement device might be less accurate due to high movement variation (Hughes et al., 2020). In the deadlift, the athletes use a barbell. The load increment, number of repetitions performed, and inter-set rest are dictated by the lifting velocity data displayed by the device's software. Table 3.1 presents a brief description of the progressive loading test in the bench press (García-Ramos, Pestaña-Melero, et al., 2018b; González-Badillo & Sánchez-Medina, 2010; Pareja-Blanco et al., 2020; Sánchez-Medina et al., 2010), back squat (Conceição

Table 3.1 Progressive Loading Test Protocols in the Bench Press, Back Squat, and Deadlift.

Bench Press	Back Squat	Deadlift
Initial position: lying on the bench in a supine position; feet resting on the bench/ground; bar grasped with a pronated grip at shoulder-width apart. Execution technique: in the eccentric phase, the barbell descends in a controlled manner (2–3 s) until it touches the chest; between the eccentric and concentric phases, there is a 1–2 s pause to avoid the rebound effect and enable more accurate measurements; concentric phase performed at the maximal intended velocity (elbows fully extended).	Initial position: upright position; barbell positioned on the upper part of the trapezius and grasped with a pronated grip; knees and hip fully extended; feet shoulder-width apart flat on the floor in parallel or externally rotated (≤ 15°). Execution technique: in the eccentric phase, the barbell descends in a controlled manner (2–3 s) until 35–45° knee flexion (full back squat) or 90° knee flexion (half back squat); between the eccentric and concentric phases, there is no pause; concentric phase performed at the maximal intended velocity (knees fully extended).	Initial position: feet shoulder-width apart flat on the floor in parallel or externally rotated (≤ 15°); natural lower back arch; barbell grasped with a closed pronated or mixed grip and elbows fully extended; chest up and head in line with the spine. Execution technique: in the concentric phase, the barbell is vertically pulled at the maximal intended velocity until reaching an upright position (knees extended just before the hip extension); the upright posture is maintained during ~1 s; in the eccentric phase, the barbell is lowered in a controlled manner to the floor; between the eccentric and concentric phases, there is a 1.5 s pause to avoid the rebound effect.
General warm-up: 5 min cycling or jogging at a self-selected pace; 5 min joint mobilization.Specific warm-up: 1–2 sets; 5–6 reps; load for men: 17–30 kg; load for women: 10 kg	General warm-up: 5 min cycling or jogging at a self-selected pace; 5 min joint mobilization and sprints.Specific warm-up: 1–3 sets; 5–6 reps; load for men: 17–40 kg; load for women: 17–20 kg	General warm-up: 5 min cycling or jogging at a self-selected pace; 5 min joint mobilization and sprints.Specific warm-up: 2 sets; 5 reps; load: 20 kg

(*Continued*)

Table 3.1 (Continued)

Bench Press	Back Squat	Deadlift
Initial load: 17–20 kgLoad increment: 10 kg for MPV ≥ 0.50 m·s^{-1}; 1–5 kg for MPV < 0.50 m·s^{-1}	Initial load: 17–20 kgLoad increment: 10 kg for MPV ≥ 0.70 m·s^{-1}; 2.5–5 kg for MPV < 0.70 m·s^{-1}	Initial load: 20 kgLoad increment: 20 kg for MPV > 0.80 m·s^{-1}; 10 kg for MPV 0.60–0.80 m·s^{-1}; 5 kg for MPV 0.50–0.60 m·s^{-1}; 2.5 kg for MPV < 0.50 m·s^{-1}
Repetitions: 3 reps for MPV > 1.00 m·s^{-1}; 2 reps for MPV 0.65–1.00 m·s^{-1}; 1 rep for MPV < 0.65 m·s^{-1}	Repetitions: 3 reps for MPV > 1.15 m·s^{-1}; 2 reps for MPV 0.70–1.15 m·s^{-1}; 1 rep for MPV < 0.70 m·s^{-1}	Repetitions: 3 reps for MPV ≥ 0.80 m·s^{-1}; 2 reps for MPV 0.60–0.80 m·s^{-1}; 1 rep for MPV < 0.60 m·s^{-1}
Inter-set rest: 2–3 min for MPV ≥ 0.50 m·s^{-1}; 5–6 min for MPV < 0.50 m·s^{-1}	Inter-set rest: 2–3 min for MPV ≥ 0.70 m·s^{-1}; 5–6 min for MPV < 0.70 m·s^{-1}	Inter-set rest: 3 min

MPV: Mean propulsive velocity.

et al., 2016; Pareja-Blanco et al., 2020; Sánchez-Medina et al., 2017), and deadlift (Benavides-Ubric et al., 2020; Morán-Navarro et al., 2020).

3.3.2 Velocity Parameters

The three velocity parameters most reported in VCRT literature are the mean velocity, mean propulsive velocity, and peak velocity. The mean velocity is the average velocity from the start of the concentric phase until the barbell reaches the maximum height. The mean propulsive velocity is the average velocity from the start of the concentric phase until the barbell acceleration is lower than the acceleration due to gravity (i.e., < −9.81 m·s^{-2}) (Sánchez-Medina et al., 2010). The peak velocity is the maximum instantaneous velocity value reached during the concentric phase.

The reliability of the three velocity variables to estimate the relative load depends on the type of exercise and relative load range. For example, during the Smith machine bench press, the mean propulsive velocity seems more accurate than the mean velocity to determine the relative load because it does not consider the braking phase's influence when using low-to-moderate loads (González-Badillo & Sánchez-Medina, 2010; Sánchez-Medina et al., 2010). Conversely, although both reliable, the mean velocity seems more accurate than the mean propulsive velocity to estimate the relative load during the Smith machine bench press throw (García-Ramos, Pestaña-Melero, et al., 2018a). When using the free-weight back squat, although all velocity variables seem reliable to estimate the relative load, only the peak velocity seems accurate to predict the 1RM load (Banyard et al., 2018). Finally, for the loaded countermovement jump, the most reliable and sensitive variable to explain and predict the vertical performance is the peak velocity (Peña García-Orea et al., 2018). Therefore, considering these differences, coaches and researchers should be encouraged to select a reference device for measuring lifting velocity first and then test which velocity variables produce the most reliable results to estimate, prescribe, and monitor the relative load.

3.3.3 Load-Velocity Data Modeling

Researchers have been modeling the load-velocity relationship by using different regression models. For example, in the Smith machine bench press exercise, some authors developed load-velocity equations by using polynomial models (González-Badillo & Sánchez-Medina, 2010; Pareja-Blanco et al., 2020), while others linear models (García-Ramos, Pestaña-Melero, et al., 2018b; Loturco et al., 2017). When comparing the reliability of both regression models to estimate the relative load in the Smith machine bench press, both models present a very high predictive ability (coefficient of determination (r^2) ~ 0.98) in two bench press variants (Pestaña-Melero et al., 2018). The same evidence remains valid for the free-weight back squat, where no differences existed between the load-velocity profiles' correlations using both models (Banyard et al., 2018). Nevertheless, surprisingly, no VCRT study analyzed how well the proposed general equations met the assumptions underpinning regression models, such as the residuals analysis. Therefore, as suggested by some authors, it is advisable to check if the model's assumptions are satisfied (Casson & Farmer, 2014). The first and easy step is creating a scatter plot between variables to analyze the line of best fit (Casson & Farmer, 2014). For example, one can suppose the data exhibit a curvilinear pattern. In that case, a quadratic model might provide the best curve fitting to the data, while if it follows a linear pattern, then a linear model might be the best option (Casson & Farmer, 2014). Therefore, instead of choosing the regression models based on previous studies' recommendations, coaches and researchers should adjust the model based on their data.

3.4 Practical Application in a Velocity-Controlled Resistance Training Setting

This section provides a practical example explaining how to integrate VCRT following specific steps using the T-Force system. According to several validation studies, the T-Force presents low measurement error and high sensitivity to detect minimal velocity changes (minimal detectable change: 0.01–0.06 m·s^{-1}) (Courel-Ibáñez et al., 2019; Martínez-Cava et al., 2020). Using a highly accurate and sensitive device is of utmost importance in VCRT settings, considering that an increase in movement velocity between 0.05 and 0.10 m·s^{-1}

against the same absolute load is related to ~5% 1RM improvements in the bench press, full back squat, and prone bench pull (González-Badillo & Sánchez-Medina, 2010; Martínez-Cava et al., 2019; Sánchez-Medina et al., 2014, 2017). In this practical example, first, we explain how to assemble the T-Force coupled with the barbell. We then present a testing example using the device's software and the statistics to model the load-velocity data. Finally, we demonstrate a training example using the T-Force software to control the training load.

3.4.1 Assembly and Calibration of the T-Force System

The T-Force system encompasses a transducer, interface, and software. The transducer consists of a solid and stable base placed on the ground or box. It does not need an external power supply, and it is connected to the computer through the interface. The latter is connected to the computer through a UBS port (González-Badillo, Sánchez-Medina, et al., 2017). Finally, the transducer cable can be attached to the barbell using a steel snap hook with a nylon cable tie. Figure 3.1 illustrates the T-Force assembly and the connection between the transducer and the barbell.

After correctly assembling the equipment, the first thing to do in the software is to ensure that the constant of calibration (K) number matches with the certified calibration number registered on the transducer. The K number is determined by a high-precision digital height gauge calibrated by the National Institute of Aerospace Technology (INTA, Spain) (González-Badillo, Sánchez-Medina, et al., 2017). After this mandatory procedure, the T-Force system is ready to be used for testing and training purposes.

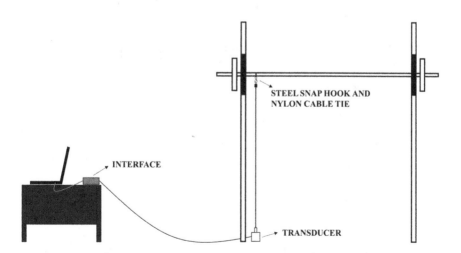

Figure 3.1 T-Force assembly.

3.4.2 Testing Procedures

The T-Force software presents a specific mode for testing purposes. After selecting the athlete, the exercise, and the test mode, the software displays in real-time a data set of the mechanical variables (i.e., velocity, force, and power) during the concentric or eccentric phases (González-Badillo, Sánchez-Medina, et al., 2017). Figure 3.2 shows the acquisition data screen during the progressive loading test in the bench press exercise.

When the test ends, the data set can be saved and stored in the T-Force software. The user can then access and analyze the data set anytime (Figure 3.3).

3.4.3 Load-Velocity Data Modeling

To model the load-velocity data obtained in the test, the user needs to export it to an Excel sheet. For example, from the data set presented in Figure 4.3, two regression equations can be modeled. One to identify the movement velocity values associated with each relative load, and a second to estimate the relative load based on the attained lifting velocity value during the training sets (Figure 3.4).

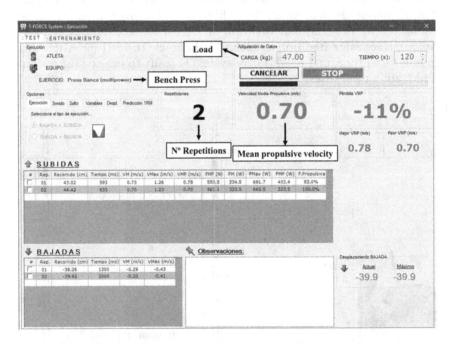

Figure 3.2 Acquisition data screen of the T-Force software during the progressive loading test in the bench press exercise.

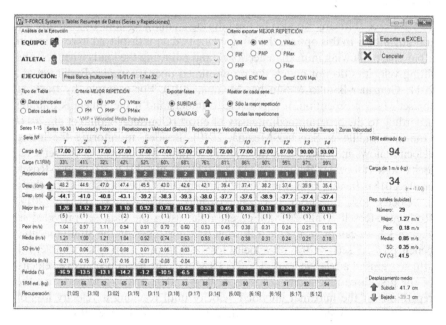

Figure 3.3 Data set summary of the progressive loading test in the bench press exercise.

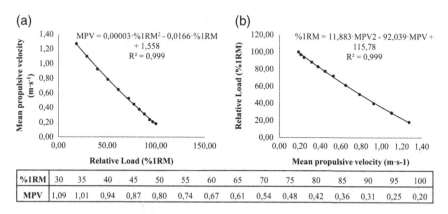

%1RM	30	35	40	45	50	55	60	65	70	75	80	85	90	95	100
MPV	1,09	1,01	0,94	0,87	0,80	0,74	0,67	0,61	0,54	0,48	0,42	0,36	0,31	0,25	0,20

Figure 3.4 Individual load-velocity relationship equations in the bench press. Equation A estimates the mean propulsive velocity (dependent variable) associated with each relative load (independent variable), as observed in the table below. Equation B estimates the relative load (dependent variable) based on the mean propulsive velocity (independent variable).

3.4.4 Training Monitoring

The T-Force software can also be used during training sessions by selecting the training mode. In this option, the lifting velocity data is displayed in real-time in a bar chart, allowing immediate feedback and adjusting the loads whenever the lifting velocities do not match the programmed ones (González-Badillo et al., 2011; González-Badillo & Sánchez-Medina, 2010). For example, by tracking the set's fastest repetition, it is possible to understand if the athlete is training according to the programmed intensity of effort (González-Badillo et al., 2011; González-Badillo & Sánchez-Medina, 2010). Besides, monitoring the lifting velocity allows individualizing the training volume by controlling the set's velocity loss (González-Badillo, Yañez-Garcia, et al., 2017; Rodríguez-Rosell et al., 2018). In this case, the athlete performs the repetitions until reaching a velocity loss threshold (González-Badillo, Yañez-Garcia, et al., 2017; Rodríguez-Rosell et al., 2018). For example, using the data set provided in Figure 3.4, we can prescribe the following training session in the Smith machine bench press: 3 sets × 20% of velocity loss × 0.54 m·s^{-1}. In this example, the athlete trains at a relative load of ~70% 1RM (i.e., the mean propulsive velocity of the first three repetitions needs to be 0.54 ± 0.03 m·s^{-1}) and performs the repetitions at the maximal intended velocity until reaching a velocity loss of ~20%. Figure 3.5 shows the acquisition data screen during VCRT using the bench press exercise.

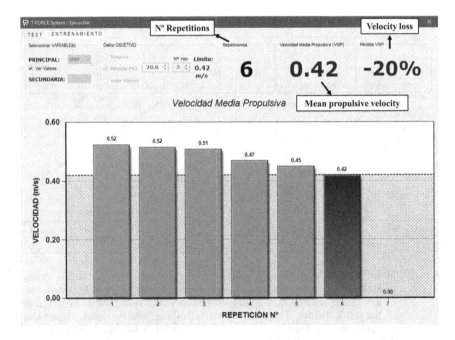

Figure 3.5 Acquisition data screen of the T-Force software during training using the bench press exercise.

3.5 Conclusion

Velocity-controlled resistance training is becoming adopted worldwide by strength and conditioning coaches due to its effectiveness in assessing, prescribing, and monitoring the training load. As a result, specialized companies continuously develop movement velocity measurement devices embedded with different technologies to ensure valid, reliable, and consistent data to coaches and researchers. Within the wide range of measurement devices, coaches should be aware of their validity and reliability for measuring movement velocity and opt for the ones that better suit their budget and training context. To further explore the benefits of measurement devices in resistance training settings, coaches should apply validated testing procedures, collect the velocity parameters according to the type of exercise, and develop load-velocity equations based on the model that provides the best curve fitting.

Acknowledgments

Portuguese Foundation for Science and Technology, I.P., funded this research, grant number SFRH/BD/147608/2019 and project UIDB/04045/2020.

References

Balsalobre-Fernández, C., Kuzdub, M., Poveda-Ortiz, P., & Campo-Vecino, J. del. (2016). Validity and Reliability of the PUSH Wearable Device to Measure Movement Velocity During the Back Squat Exercise. *The Journal of Strength & Conditioning Research*, *30*(7). 1968–1974. 10.1519/JSC.0000000000001284

Balsalobre-Fernández, C., Marchante, D., Baz-Valle, E., Alonso-Molero, I., Jiménez, S. L., & Muñóz-López, M. (2017). Analysis of Wearable and Smartphone-Based Technologies for the Measurement of Barbell Velocity in Different Resistance Training Exercises. *Frontiers in Physiology*, *8*, 649. 10.3389/fphys.2017.00649

Balsalobre-Fernández, C., Marchante, D., Muñoz-López, M., & Jiménez, S. L. (2018). Validity and Reliability of a Novel iPhone App for the Measurement of Barbell Velocity and 1RM on the Bench-Press Exercise. *Journal of Sports Sciences*, *36*(1), 64–70. 10.1080/02640414.2017.1280610

Banyard, H. G., Nosaka, K., Sato, K., & Haff, G. G. (2017). Validity of Various Methods for Determining Velocity, Force, and Power in the Back Squat. *International Journal of Sports Physiology and Performance*, *12*(9), 1170–1176. 10.1123/ijspp.2016-0627

Banyard, H. G., Nosaka, K., Vernon, A. D., & Haff, G. G. (2018). The Reliability of Individualized Load–Velocity Profiles. *International Journal of Sports Physiology and Performance*, *13*(6), 763–769. 10.1123/ijspp.2017-0610

Benavides-Ubric, A., Díez-Fernández, D. M., Rodríguez-Pérez, M. A., Ortega-Becerra, M., & Pareja-Blanco, F. (2020). Analysis of the Load-Velocity Relationship in Deadlift Exercise. *Journal of Sports Science & Medicine*, *19*(3), 452–459. https://pubmed.ncbi.nlm.nih.gov/32874097

Bosquet, L., Porta-Benache, J., & Blais, J. (2010). Validity of a Commercial Linear Encoder to Estimate Bench Press 1 RM from the Force-Velocity Relationship. *Journal of Sports Science & Medicine, 9*(3), 459–463. https://pubmed.ncbi.nlm.nih.gov/24149641

Casson, R. J., & Farmer, L. D. M. (2014). Understanding and Checking the Assumptions of Linear Regression: A Primer for Medical Researchers. *Clinical & Experimental Ophthalmology, 42*(6), 590–596. 10.1111/ceo.12358

Chéry, C., & Ruf, L. (2019). Reliability of the Load-Velocity Relationship and Validity of the PUSH to Measure Velocity in the Deadlift. *The Journal of Strength & Conditioning Research, 33*(9). 2370–2380. 10.1519/JSC.0000000000002663

Conceição, F., Fernandes, J., Lewis, M., González-Badillo, J. J., & Jímenez-Reyes, P. (2016). Movement Velocity as a Measure of Exercise Intensity in Three Lower Limb Exercises. *J Sports Sci, 34*(12), 1099–1106. 10.1080/02640414.2015.1090010

Courel-Ibáñez, J., Martínez-Cava, A., Morán-Navarro, R., Escribano-Peñas, P., Chavarren-Cabrero, J., González-Badillo, J. J., & Pallarés, J. G. (2019). Reproducibility and Repeatability of Five Different Technologies for Bar Velocity Measurement in Resistance Training. *Annals of Biomedical Engineering, 47*(7), 1523–1538. 10.1007/s10439-019-02265-6

de Sá, E. C., Ricarte Medeiros, A., Santana Ferreira, A., García Ramos, A., Janicijevic, D., & Boullosa, D. (2019). Validity of the iLOAD® App for Resistance Training Monitoring. *PeerJ, 7*, e7372, 1–14. 10.7717/peerj.7372

García-Ramos, A., Pérez-Castilla, A., & Martín, F. (2018). Reliability and Concurrent Validity of the Velowin Optoelectronic System to Measure Movement Velocity during the Free-Weight Back Squat. *International Journal of Sports Science & Coaching, 13*(5), 737–742. 10.1177/1747954118791525

García-Ramos, A., Pestaña-Melero, F. L., Pérez-Castilla, A., Rojas, F. J., & Gregory Haff, G. (2018a). Mean Velocity vs. Mean Propulsive Velocity vs. Peak Velocity: Which Variable Determines Bench Press Relative Load With Higher Reliability? *The Journal of Strength & Conditioning Research, 32*(5). 10.1519/JSC.0000000000001998

García-Ramos, A., Pestaña-Melero, F. L., Pérez-Castilla, A., Rojas, F. J., & Haff, G. G. (2018b). Differences in the Load–Velocity Profile between 4 Bench-Press Variants. *International Journal of Sports Physiology and Performance, 13*(3), 326–331. 10.1123/ijspp.2017-0158

Garnacho-Castaño, M. V., López-Lastra, S., & Maté-Muñoz, J. L. (2015). Reliability and Validity Assessment of a Linear Position Transducer. *Journal of Sports Science & Medicine, 14*(1), 128. https://pubmed.ncbi.nlm.nih.gov/25729300/

Goldsmith, J. A., Trepeck, C., Halle, J. L., Mendez, K. M., Klemp, A., Cooke, D. M., Haischer, M. H., Byrnes, R. K., Zoeller, R. F., Whitehurst, M., & Zourdos, M. C. (2019). Validity of the Open Barbell and Tendo Weightlifting Analyzer Systems Versus the Optotrak Certus 3D Motion-Capture System for Barbell Velocity. *International Journal of Sports Physiology and Performance, 14*(4), 540–543. 10.1123/ijspp.2018-0684

González-Badillo, J. J., Marques, M. C., & Sánchez-Medina, L. (2011). The Importance of Movement Velocity as a Measure to Control Resistance Training Intensity. *J Hum Kinet, 29A*, 15–19. 10.2478/v10078-011-0053-6

González-Badillo, J. J., & Sánchez-Medina, L. (2010). Movement Velocity as a Measure of Loading Intensity in Resistance Training. *Int J Sports Med*, *31*(05), 347–352. 10.1055/s-0030-1248333

González-Badillo, J. J., Sánchez-Medina, L., Pareja-Blanco, F., & Rodríguez-Rosell, D. (2017). *La velocidad de ejecución como referencia para la programación, control y evaluación del entrenamiento de fuerza* (1st ed.). ERGOTECH Consulting, S L.

González-Badillo, J. J., Yañez-Garcia, J. M., Mora-Custodio, R., & Rodríguez-Rosell, D. (2017). Velocity Loss as a Variable for Monitoring Resistance Exercise. *Int J Sports Med*, *38*(3), 217–225. 10.1055/s-0042-120324

Grgic, J., Scapec, B., Pedisic, Z., & Mikulic, P. (2020). Test-Retest Reliability of Velocity and Power in the Deadlift and Squat Exercises Assessed by the GymAware PowerTool System. *Frontiers in Physiology*, *11*, 1146. 10.3389/fphys.2020.561682

Hernández-Belmonte, A., Martínez-Cava, A., Morán-Navarro, R., Courel-Ibáñez, J., & Pallarés, J. (2020). A Comprehensive Analysis of the Velocity-Based Method in the Shoulder Press Exercise: Stability of the Load-velocity Relationship and Sticking Region Parameters. *Biology of Sport*, *38*(2), 235–243. 10.5114/biolsport.2020.98453

Hughes, L. J., Peiffer, J. J., & Scott, B. R. (2020). Load–Velocity Relationship 1RM Predictions: A Comparison of Smith Machine and Free-Weight Exercise. *Journal of Sports Sciences*, *38*(22), 2562–2568. 10.1080/02640414.2020.1794235

Kasovic, J., Martin, B., Carzoli, J. P., Zourdos, M. C., & Fahs, C. A. (2020). Agreement between the Iron Path App and a Linear Position Transducer for Measuring Average Concentric Velocity and Range of Motion of Barbell Exercises. *The Journal of Strength & Conditioning Research, Publish Ah*, 35, S95–S101. 10.1519/JSC.0000000000003574

Laza-Cagigas, R., Goss-Sampson, M., Larumbe-Zabala, E., Termkolli, L., & Naclerio, F. (2019). Validity and Reliability of a Novel Optoelectronic Device to Measure Movement Velocity, Force and Power during the Back Squat Exercise. *Journal of Sports Sciences*, *37*(7), 795–802. 10.1080/02640414.2018.1527673

Lorenzetti, S., Lamparter, T., & Lüthy, F. (2017). Validity and Reliability of Simple Measurement Device to Assess the Vlocity of the Barbell during Squats. *BMC Research Notes*, *10*(1), 707. 10.1186/s13104-017-3012-z

Loturco, I., Kobal, R., Moraes, J. E., Kitamura, K., Cal Abad, C. C., Pereira, L. A., & Nakamura, F. Y. (2017). Predicting the Maximum Dynamic Strength in Bench Press: The High Precision of the Bar Velocity Approach. *The Journal of Strength & Conditioning Research*, *31*(4). 1127–1131. 10.1519/JSC.0000000000001670

Marinho, D. A., & Neiva, H. P. (2020). Introductory Chapter: Rising Interests in Sports Sciences. In *Sports Science and Human Health - Different Approaches* (p. 3). IntechOpen. 10.5772/intechopen.94837

Martínez-Cava, A., Hernández-Belmonte, A., Courel-Ibáñez, J., Morán-Navarro, R., González-Badillo, J. J., & Pallarés, J. G. (2020). Reliability of Technologies to Measure the Barbell Velocity: Implications for Monitoring Resistance Training. *PLOS ONE*, *15*(6), e0232465, 1–17. 10.1371/journal.pone.0232465

Martínez-Cava, A., Morán-Navarro, R., Sánchez-Medina, L., González-Badillo, J. J., & Pallarés, J. G. (2019). Velocity- and Power-Load Relationships in the Half, Parallel and Full Back Squat. *Journal of Sports Sciences*, *37*(10), 1088–1096. 10.1080/02640414.2018.1544187

Morán-Navarro, R., Martínez-Cava, A., Escribano-Peñas, P., & Courel-Ibáñez, J. (2020). Load-Velocity Relationship of the Deadlift Exercise. *European Journal of Sport Science*, 1–7. 10.1080/17461391.2020.1785017

Mosey, T., Watts, D., Giblin, G., & Brown, W. (2018). Reliability and Validity of the "Elite Form" 3D Motion Capturing System. *Journal of Australian Strength and Conditioning*, *26*(5), 6–14. https://strengthandconditioning.org/jasc-26-5

Muniz-Pardos, B., Lozano-Berges, G., Marin-Puyalto, J., Gonzalez-Agüero, A., Vicente-Rodriguez, G., Casajus, J. A., & Garatachea, N. (2019). Validity and Reliability of an Optoelectronic System to Measure Movement Velocity during Bench Press and Half Squat in a Smith Machine. *Proceedings of the Institution of Mechanical Engineers, Part P: Journal of Sports Engineering and Technology*, *234*(1), 88–97. 10.1177/1754337119872418

Orange, S. T., Metcalfe, J. W., Liefeith, A., Marshall, P., Madden, L. A., Fewster, C. R., & Vince, R. V. (2019). Validity and Reliability of a Wearable Inertial Sensor to Measure Velocity and Power in the Back Squat and Bench Press. *The Journal of Strength & Conditioning Research*, *33*(9), 2398–2408. 10.1519/JSC.0000000000002574

Orange, S. T., Metcalfe, J. W., Marshall, P., Vince, R. V., Madden, L. A., & Liefeith, A. (2020). Test-Retest Reliability of a Commercial Linear Position Transducer (GymAware PowerTool) to Measure Velocity and Power in the Back Squat and Bench Press. *The Journal of Strength & Conditioning Research*, *34*(3), 728–737. 10.1519/JSC.0000000000002715

Pareja-Blanco, F., Rodríguez-Rosell, D., Sánchez-Medina, L., Sanchis-Moysi, J., Dorado, C., Mora-Custodio, R., Yáñez-García, J. M., Morales-Alamo, D., Pérez-Suárez, I., Calbet, J. A. L., & González-Badillo, J. J. (2017). Effects of Velocity Loss during Resistance Training on Athletic Performance, Strength Gains and Muscle Adaptations. *Scand J Med Sci Sports*, *27*(7), 724–735. 10.1111/sms.12678

Pareja-Blanco, F., Walker, S., & Häkkinen, K. (2020). Validity of Using Velocity to Estimate Intensity in Resistance Exercises in Men and Women. *Int J Sports Med*, *EFirst*, *41*(14), 1047–1055. 10.1055/a-1171-2287

Peart, D. J., Balsalobre-Fernández, C., & Shaw, M. P. (2019). Use of Mobile Applications to Collect Data in Sport, Health, and Exercise Science: A Narrative Review. *The Journal of Strength & Conditioning Research*, *33*(4), 1167–1177. 10.1519/JSC.0000000000002344

Peña García-Orea, G., Belando-Pedreño, N., Merino-Barrero, J. A., Jiménez-Ruiz, A., & Heredia-Elvar, J. R. (2018). Validation of an Opto-Electronic Instrument for the Measurement of Weighted Countermovement Jump Execution Velocity. *Sports Biomechanics*, *20*(2), 150–164. 10.1080/14763141.2018.1526316

Pérez-Castilla, A., Piepoli, A., Delgado-García, G., Garrido-Blanca, G., & García-Ramos, A. (2019). Reliability and Concurrent Validity of Seven Commercially Available Devices for the Assessment of Movement Velocity at Different Intensities During the Bench Press. *The Journal of Strength & Conditioning Research*, *33*(5). 10.1519/JSC.0000000000003118

Pérez-Castilla, A., Piepoli, A., Garrido-Blanca, G., Delgado-García, G., Balsalobre-Fernández, C., & García-Ramos, A. (2019). Precision of 7 Commercially Available Devices for Predicting Bench-Press 1-Repetition Maximum from the Individual Load–Velocity Relationship. *International Journal of Sports Physiology and Performance*, *14*(10), 1442–1446. 10.1123/ijspp.2018-0801

Pestaña-Melero, F. L., Haff, G. G., Rojas, F. J., Pérez-Castilla, A., & García-Ramos, A. (2018). Reliability of the Load–Velocity Relationship Obtained Through Linear and Polynomial Regression Models to Predict the 1-Repetition Maximum Load. *Journal of Applied Biomechanics, 34*(3), 184–190. 10.1123/jab.2017-0266

Rodríguez-Rosell, D., Yáñez-García, J. M., Mora-Custodio, R., Pareja-Blanco, F., Ravelo-García, A. G., Ribas-Serna, J., González-Badillo, J. J., Yanez-Garcia, J. M., Mora-Custodio, R., Pareja-Blanco, F., Ravelo-Garcia, A. G., Ribas-Serna, J., & Gonzalez-Badillo, J. J. (2020). Velocity-Based Resistance Training: Impact of Velocity Loss in the Set on Neuromuscular Performance and Hormonal Response. *Appl Physiol Nutr Metab*. 10.1139/apnm-2019-0829

Rodríguez-Rosell, D., Yanez-Garcia, J. M., Torres-Torrelo, J., Mora-Custodio, R., Marques, M. C., & González-Badillo, J. J. (2018). Effort Index as a Novel Variable for Monitoring the Level of Effort during Resistance Exercises. *J Strength Cond Res, 32*(8), 2139–2153. 10.1519/jsc.0000000000002629

Sánchez-Medina, L., & González-Badillo, J. (2011). Velocity Loss as an Indicator of Neuromuscular Fatigue during Resistance Training. *Med Sci Sports Exerc, 43*(9), 1725–1734. 10.1249/MSS.0b013e318213f880

Sánchez-Medina, L., González-Badillo, J. J., Perez, C. E., & Pallares, J. G. (2014). Velocity- and Power-Load Relationships of the Bench Pull vs. Bench Press Exercises. *Int J Sports Med, 35*(3), 209–216. 10.1055/s-0033-1351252

Sánchez-Medina, L., Pallarés, J. G., Pérez, C. E., Morán-Navarro, R., & González-Badillo, J. J. (2017). Estimation of Relative Load from Bar Velocity in the Full Back Squat Exercise. *Sports Medicine International Open, 1*(2), E80–E88. 10.1055/s-0043-102933

Sánchez-Medina, L., Pérez, C. E., & González-Badillo, J. J. (2010). Importance of the Propulsive Phase in Strength Assessment. *International Journal of Sports Medicine, 31*(2), 123–129. 10.1055/s-0029-1242815

Sánchez-Pay, A., Courel-Ibáñez, J., Martínez-Cava, A., Conesa-Ros, E., Morán-Navarro, R., & Pallarés, J. G. (2019). Is the High-Speed Camera-Based Method a Plausible Option for bar Velocity Assessment during Resistance Training? *Measurement, 137*, 355–361. 10.1016/j.measurement.2019.01.006

Sato, K., Beckham, G., Carroll, K., Bazyler, C., Sha, A., & Haff, G. (2015). Validity of Wireless Device Measuring Velocity of Resistance Exercises. *J. Trainology, 4*(1), 15–18. 10.17338/trainology.4.1_15

Thompson, S. W., Rogerson, D., Dorrell, H. F., Ruddock, A., & Barnes, A. (2020). The Reliability and Validity of Current Technologies for Measuring Barbell Velocity in the Free-Weight Back Squat and Power Clean. *Sports, 8*(7), 94. 10.3390/sports8070094

Tomasevicz, C. L., Hasenkamp, R. M., Ridenour, D. T., & Bach, C. W. (2020). Validity and Reliability Assessment of 3-D Camera-Based Capture Barbell Velocity Tracking Device. *Journal of Science and Medicine in Sport, 23*(1), 7–14. 10.1016/j.jsams.2019.07.014

4 Deliberate Practice, Sports Expertise, and Instincts That Can Be Taught Using Interactive Sports Technologies

Peter J. Fadde and Scott Boatright

A point-guard skips a no-look pass between defenders; a linebacker shoots the gap to blow up a screen pass; a baseball batter laces a slider to the opposite field. Invariably, a TV announcer declares, "You can't coach instincts like that!" And we are thrown into the long-standing nature-versus-nurture, trait-versus-state, talent-versus-practice debate. In recent years, researchers in the academic cross-disciplinary field of Expertise Studies (combining Cognitive Psychology, Human Factors, and Learning Sciences) have swung the debate to the nurture/state/practice nexus. In particular, expertise studies introduced the concept of *deliberate practice*, which has been popularized as the 10,000-hour Rule (Gladwell, 2008). As noted expertise researcher Anders Ericsson wrote in *Peak: Secrets from the New Science of Expertise* (Ericsson & Pool, 2016), the primary secret is that expertise and expert performance come more from sustained and strategic (i.e., deliberate) practice more than from in-born talent.

In the context of sports, *sports expertise* researchers have focused on cognitive rather than physical, technical, or psychological components of expert sports performance. Specifically, sports expertise researchers have looked at skills in open play sports such as tennis, baseball, cricket, football, and hockey that involve fast reactions. Expert performers in fast reaction sports are able to "read" the actions of an opponent – whether that is a tennis server or a whitewater course – faster and better than lesser skilled performers. The implications are that this expert advantage is both mea-sureable and trainable, and that interactive sports technologies have a key role (Fadde & Zaichkowsky, 2019).

This chapter challenges the assumption that the primary goal of simulator designs, for the purpose of improving athletic performance, is to capture real-life experience as realistically as possible. Unfortunately, design of simulators and virtual reality environments is almost entirely addressed as an engineering challenge with inadequate consideration of the learning challenge. As a result, technologies such as video-based temporal occlusion that are validated by decades of research (Müller, Fitzgerald, & Brenton, 2020) remain under-developed as commercial products and almost unknown in applied sports

DOI: 10.4324/9781003205111-4

training contexts. There is, therefore, both a need and an opportunity for interactive sports technologies that can target these skills in less expensive, more flexible, more efficient and, ultimately, more effective ways.

Because our premise that "more realistic doesn't necessarily mean more learning" is counter-intuitive, we start with a deep dive into the theoretical foundations for training decision and perception components of fast-reaction sports. We look at the new academic field of Expertise Studies, in which classic chess studies provide a framework for sports expertise. We consider the role of Deliberate Practice in training sports expertise, and offer a framework for categorizing sports skills as Open (reactive) or Closed (self-paced) and rely more on Athleticism or on Technique (see Figure 4.1). Our primary interest is in open, technique-based sports skills where elite performers seem to have an instinctive sense of what will happen next. With a foundation of understanding the targeted skills, we then ask applied questions: Are sports instincts real? What do these skills entail? How can they be measured and trained? What are the benefits and opportunities for research and technology development?

While the vast majority of sports skills training focuses on "what you do" it's widely acknowledged that high levels of performance increasingly depend on "what you see" and are able to turn into rapid psychomotor actions

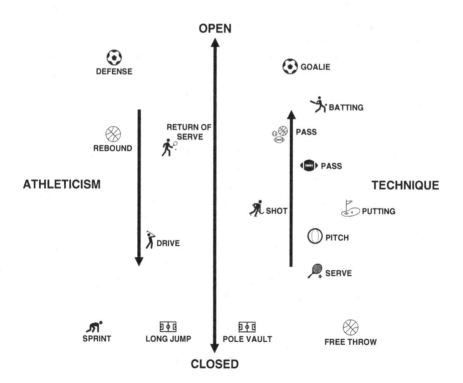

Figure 4.1 Sports skills as Open or Closed and favoring Athleticism or Technique.

(Zaichkowsky & Peterson, 2018). Not only does sports expertise research provide supporting theories and findings but also has developed novel interactive sports technologies in order to measure and, ultimately, improve expert decision and perception skills. These interactive technologies that were first developed as research tools, such as video-occlusion, provide alternatives to virtual reality and simulator-based training that potentially can be less expensive and more effective.

4.1 Expertise Studies

To understand sports expertise as an academic area, we need to view it in the framework of expertise studies. While academic disciplines are too multifaceted to credit their origin to a single experiment or publication, expertise studies has clearly identifiable headwaters in early 1970s chess studies (e.g., Chase & Simon, 1973). Chase and Simon actually were not concerned with improving performance in chess. Rather, they were studying short-term memory in order to advance artificial intelligence. Chess offered a memory intensive skill and they pioneered the *expert-novice* research paradigm (commonly used in expertise studies) to capture expert performance. Chase and Simon's experiments found that an internationally ranked chess master could complete a chess memory task better than an accomplished, but not ranked, chess player could – not a surprising finding, except that it came with a key condition.

Chase and Simon ran two experiments. When the chess players were given a short period of time to look at a chessboard with many pieces set on it, and the pieces reproduced a legal game arrangement, then the expert excelled at remembering the placement of pieces. But when the chess pieces were arranged arbitrarily on the chessboard, the expert's advantage almost completely disappeared. Chase and Simon (1973) speculated that experts *chunk* information. For instance, chess players may notice that the positions of the rook and king on the back row permitted a castling maneuver, and then store this as one piece of information. Later, the expert can unpack that one chunk of information to place several chess pieces. Chunking let experts circumvent the then accepted 7-Plus-or Minus-2 limit for short-term memory (Miller, 1956). While mastering the complex game of chess obviously requires high levels of memory and cognition, the expert in Chase and Simon's experiment enjoyed a software advantage (chess-specific schema) rather than a hardware advantage (prodigious memory). In the closing discussion section of their seminal article, Chase and Simon speculated, rather informally, what years later has become known as the 10,000-Hour Rule:

> *There appears not to be any case (including Bobby Fischer) where a person has reached grandmaster level with less than about a decade's intense preoccupation with the game. We would estimate, very roughly, that a master has spent perhaps 10,000 to 50,000 hours staring at chess positions, and a Class A player 1,000 to 5,000 hours.*

In 1993, 20 years after the chess studies of Chase and Simon, Anders Ericsson (Chase's former doctoral student) and colleagues elaborated on the 10,000-hour rule in the domain of music. In *The Role of Deliberate Practice in the Acquisition of Expert Performance* they compared numerous characteristics of students at a highly esteemed music school. The researchers compared groups of top students (as identified by faculty), less advanced students, professional violinists, and teachers. The best predictor of performance category was the amount of *deliberate* practice the music students had engaged in. Simply, the most advanced students had practiced substantially more than the accomplished but less advanced students. Professional concert musicians had, at a similar point in their development, practiced more hours than those that became music teachers. The average hours of practice put in by advanced students and professional musicians (at that point in their development) approached 10,000 hours (Ericsson, Krampe, & Tesch-Römer, 1993). The 10,000-hour rule has come to represent the predominance of deliberate practice over talent in developing expertise. Benjamin Bloom's *Developing Talent in Young People* (1985) similarly reports early and intense practice by expert performers in domains including mathematics, music, art, ballet, and sports. Overall, expertise studies stress that what appears to be innate talent is often the result of an early start, family support, and increasingly elite coaching featuring deliberate practice (Ericsson, 2003, 2006).

4.2 Training Sports Expertise

While acknowledging that there are many dimensions of athletic performance, including strength and conditioning, nutrition, skill acquisition, and psychological skills, our interest is the type of in-the-action decision-making that appears to observers (and often to performers themselves) to be intuitive or instinctive (Furley, Schweizer, & Bertrams, 2015). Are sports instincts real? How can they be measured? How can they be trained? Sports expertise researchers focus on these questions as the essence of expertise in reactive sports.

Figure 4.1 places a variety of sports skills along two dimensions. One dimension is the degree to which the skills are *open* or *closed*. In general, team invasion sports such as soccer, basketball, hockey, and American football are categorized as open sports although individual skills within these sports, such as a basketball free throw, are closed skills. By contrast, individual sports against an opponent, such as boxing, tennis, and badminton, are open sports but again with closed skills, such as the serve in racquet sports. Although more of a continuum than set categories, closed sports contested against a clock, such as track-and-field and swimming, measurements of distance jumped or thrown, or practiced routines that are judged, such as diving, gymnastics, and figure skating. Most closed sports skills are also internally or self-paced while open sports are externally paced.

A second dimension we track is the degree to which the skills emphasize *athleticism* or *technique*. Though almost all sports combine athleticism and technique, they value them differently. Pole-vaulting for instance, certainly involves substantial athleticism but, compared to sprinting, also requires mastery of challenging techniques. On the other hand, free-throw shooting in basketball is a completely closed skill within a free-flowing open sport. Putting in golf essentially is a closed and technique intensive skill, but requires adapting actions to environmental conditions. Driving a golf ball does not require less technique than putting, but does reward greater athleticism.

Sports skills that are categorized as being open or closed can also be categorized as being externally or internally paced. Almost all open skills in which athletes react to the actions of an opponent are externally paced. However, ball distribution in team invasion sports such as soccer, basketball, and hockey is an open skill that is internally paced. The ball handler decides when and where to pass the ball or puck. Ball distribution in American football is more closed than in other invasion sports because the quarterback usually decides between defined passing options in a choreographed play.

While sports are often described as open or closed, specific skills within a sport – sometimes related to playing position – can be more or less open than their parent sport. Tennis, for instance, is considered an open sport but the most important shot (the serve) is essentially a closed, internally paced skill that tennis players can practice without a coach, an opponent, or even a court. For that reason, serving tends to be practiced much more often and more systematically than the second most important shot in tennis – return of serve (Williams, Ward, Smeeton, & Allen, 2004). As an open, externally paced skill, return-of-serve requires an opponent and a court, and is therefore assumed primarily to develop during competition more than during practice. The framework of *deliberate practice*, however, does not count competition play in the 10,000 hours. Deliberate practice tends to occur under the guidance of a coach and is intended to address deficiencies or in measurable ways that can be repeated and refined (Ericsson, Krampe, & Tesch-Römer, 1993).

In other words, "play is not practice" – at least not deliberate practice. Indeed, one of the criticisms of modern youth sports is that there are too many games and not enough team practice. Developing athletes typically receive highly focused individual training on closed skills, often with private coaches, but then compete on traveling or tournament teams that offer high-level competition but usually practice very little. In particular, showcase events for high-performance youth in many sports offer developing athletes a vehicle to *prove* their performance more than to *improve* their performance.

Categorizing sports skills according to whether they are open or closed, and internally or externally paced, has implications for the design of deliberate practice. Closed skills primarily involve *production* (what you do) and can therefore be practiced without an opponent. Open skills have a production component but also involve *recognition* (what you see) where the choice and execution of movements are based on reading changes in the environment, which could include shadows on a downhill ski run as well as groundstroke tendencies of an opponent across the net. Practicing open skills, therefore, requires an opponent and gameplay context. For both closed and open sports skills, initial skill acquisition focuses deliberate practice on consistent production of psychomotor actions. For open sport skills, deliberate practice in more advanced stages of an athlete's development expand to include recognition of opponents' actions.

The emphasis of Ericsson and others on deliberate practice over innate talent or physical attributes is not uncontroversial, with David Epstein arguing in *The Sports Gene* that inherited characteristics can facilitate sports expertise – but noting that these facilitative characteristics include not only size and speed but also things such as pain tolerance. Still, examples abound of athletes who reach the highest levels of competition in their sports without appearing to possess prodigious talent or physical attributes. Although Epstein maintains that inborn attributes may enable some athletes to require less practice to attain expertise, the indisputable conclusion remains that a very substantial commitment to thousands of hours of deliberate practice is necessary – if not sufficient – to achieve the highest levels of performance in virtually any domain of performance, including sports.

Clearly, different types of sports skills call for different amounts and types of deliberate practice. Skills that are closed and athletic require a large commitment to training of sport-specific strength and conditioning. Skills that are closed, technical, and internally paced require concentrated, part-task practice to perfect production skills. Skills that are open and athletic require concentrated practice with opponents. Open skills that rely on opponent recognition to select responses are typically assumed to require whole-task simulation or competition match play. However, if the recognition component of reactive sports skills can be trained in targeted part-task ways, similar to closed skills, then the efficiency and volume of training can be substantially increased. Sports expertise researchers have investigated the mechanisms of seemingly instinctive skills, asking whether these skills are real, what exactly they entail, how they can be measured and, ultimately, how they can be improved through deliberate practice.

4.3 Are Sports Instincts Real?

Athletes in many sports display rapid decision making that is executed extremely quickly and with almost no conscious effort, thereby appearing to be intuitive (Furley, Schweizer, & Bertrams, 2015). But they were not born with these skills. Nor are these skills generalizable. They are specific to a sport and even to particular positions (such as point-guard or goalie) and skills within a sport. Indeed, as described by David Epstein in *The Sports Gene*, the very best major league baseball hitters of a generation (Albert Pujols and Barry Bonds) were not able to hit softballs pitches thrown by professional softball pitcher, Jennie Finch. Epstein's conclusion, like that of sports expertise researchers, is that expert batters rely on pattern-recognition (software) in addition to vision and physical attributes (hardware). Pujols and Bonds had the visual and physical hardware to hit Finch's pitches, but not the softball-specific software.

4.4 What Do These Skills Entail?

Sports expertise researchers look at two skill types: *perceptual-cognitive* and *decision* skills. Perceptual-cognitive skills are ballistically reactive, often requiring recognition, response selection, and response execution in less than one second (Williams, 2020). A major league baseball pitch, for instance, takes about 400 milliseconds to travel from the pitcher's hand to the batter's hitting zone. It takes batters at least 150 milliseconds to execute their swings, affording batters approximately 250 milliseconds (literally, an eye blink) to decide whether and where to swing their bats. And yet the best batters routinely hit pitches that are not only traveling rapidly but also moving in unpredictable ways. Returning 130 mile-per-hour tennis serves and blocking shots on goal in hockey and soccer are other examples of perceptual-cognitive skills.

In contrast to perceptual-cognitive skills that are executed in less than a second, *decision* skills – typified by the ball distribution decisions of a quarterback, mid-fielder, or point guard – afford the performer a generous two or three seconds to recognize the situation, choose a response, and execute the action (Williams, 2020). However, these decision skills are still considered intuitive, meaning that they are an automatic process that can be practiced (Furley, Schweizer, & Bertrams, 2015).

4.5 How Can Perceptual-Cognitive and Decision Skills Be Measured?

Because perception, decision, and action are closely linked in reactive sports it is difficult to disentangle them for research or training purposes. Was a "no-look" pass by Wayne Gretzky or Magic Johnson the result of strategic pre-game study? Or was it unconscious pattern recognition? Or was it vision and eye-hand coordination?

Sports expertise researchers, then, needed to isolate the perceptual-cognitive and decision components away from psychomotor performance in order to verify and quantify these component skills (Roca & Williams, 2016). To do this, sports expertise researchers adopted the *expert-novice* framework and *representative task* methods that were first developed in chess studies and became commonly used in expertise studies.

Typical sports expertise experiments in a laboratory involved comparing highly skilled performers to lesser skilled performers while completing a task that captures key elements of on-field/court/ice performance. For example, a laboratory experiment involved tennis players standing with their feet on force pads and watching video clips of a server that were recorded from the point-of-view of a service receiver. The participants reacted to the video serves by stepping to backhand or forehand side (and stepping on the force pad) while calling out loud "flat," "slice," or "kick" to identify the type of serve (Williams et al., 2004).

Knowing that returners need a certain amount of time to position their racquets in order to return high-speed serves makes it imperative that returners pick up advance and early cues to anticipate serve type and placement to backhand, forehand, or body (Williams et al., 2004). Therefore, researchers created *occlusion* video clips in which different visual information are removed (Fadde, 2006). If an experimental condition resulted in loss of expert advantage (as with non-match arrangement of pieces in the chess studies) then researchers assumed that the occluded information was important to the experts' superior performance. In *spatial occlusion* the server's racquet may be masked. If expert tennis players are better than novice players at identifying the type of serve even without seeing the racquet-ball contact, then expert receivers must pick up advance cues from other sources, such as the server's knee bend, shoulder rotation, and ball toss. On the other hand, if masking the server's legs reduces the experts' advantage, then the experts' anticipation is assumed to rely on the server's large body movements (e.g., amount of knee bend).

Experimenters sometimes verify spatial occlusion studies using eye tracking to observe where experts' look. Typically, experts have fewer fixations of longer duration and concentrate on areas of the opponent's motion that provide the best advance cues (Abernethy, 1988). For instance, expert baseball batters tend to focus on the expected release point of a pitch rather than following the pitcher's hand through its backswing motion. Indeed, across many sports, research shows the experts often *don't* keep their eye on the ball!

Another type of occlusion is *temporal occlusion* where the opponent's action is cut off in time. Typically, point-of-view video of an opponent server or pitcher is edited to black at different points during the opponent's motion or early ball flight. Expert and novice performers are tasked with identifying the type of serve or pitch. Typically, if more than a certain amount of ball flight is shown then the performance of experts and novices is equally good. On the other hand, if less than a certain amount of the opponent's action or ball flight is shown then both experts and novices are reduced to chance level in identifying serve or pitch type. The serve or pitch identification performance of both experts and novices naturally decreases as less ball flight is shown. However, the *difference* between the performance of experts and novices is greater when less visual information is available due to occlusion conditions (Fadde, 2006).

Occlusion-based expert-novice research has verified that experts pick up advance cues that underlie their seemingly impossible reaction times. In addition, studies of skills such as return-of-serve in tennis, batting in baseball, softball, and cricket, goalie play in soccer and hockey, and ball distribution in basketball and soccer have identified both where and when experts look to glean predictive information. While a very substantial body of research, over decades, has validated and explored the perceptual-cognitive skills of elite performers, a smaller body of research as validated using the same temporal and spatial occlusion research methods for training purposes. Figure 4.2 shows a commercially developed video-simulation training program in which baseball batters view video clips of opponent pitchers. Video-simulation provides a visual stimulus, accepts users' decision input (e.g., Curveball/Strike), and provides immediate feedback and a score. It is not a full simulation because users' decisions do not generate a further reaction in the simulation.

A fully immersive simulation reacts to the action input of a user. Figure 4.3 shows a virtual reality (VR) system in which a user swings a custom bat in order to "hit" the pitch. The visual display in VR is computer generated in order to be responsive to the user actions. It also allows the trajectory of specific pitchers' pitches to be accurately modeled based on radar data. On the other hand, video-simulation (see Figure 4.2) displays photo-realistic video of a pitcher rather than a 3D avatar of a pitcher. Video-based perceptual-cognitive training has been consistently validated in research settings (Larkin, Mesagno, Spittle, & Berry, 2015) and virtual reality simulation is increasingly developed, researched, and modeled (Gray, 2019). It may be that video-simulation using photo-realistic video is optimal for skill acquisition stage while immersive and customizable virtual reality is optimal for opponent preparation. However, comparative learning benefits of video-simulation and immersive virtual reality are almost entirely unexplored.

4.6 How Can Perceptual-Cognitive and Decision Skills Be Trained?

Of course, training perceptual-cognitive and decision skills in sports isn't new (Fadde & Zaichkowsky, 2019). Football coaches often end practice sessions

(a)   (b)

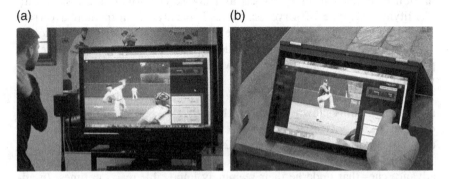

Figure 4.2 Video-simulation using temporal occlusion in the sport of baseball (photo courtesy of gameSense Sports, Inc.).

Figure 4.3 Virtual reality baseball hitting simulation (photo courtesy of WIN
Reality, LLC).

with simulations of game situations. Professional baseball players see thousands of pitches as they progress through levels of minor leagues playing
games that are primarily intended for player development. But these are all
whole-task activities. While coaches traditionally break psychomotor skills
into sub-skills in order to gain instructional efficiencies, perceptual-cognitive
and decision skills are often assumed to require simulated or actual match
performance. The simple but significant insight of sports expertise research is
that the development of perceptual-cognitive sub-skills such as baseball pitch
recognition can be accelerated by repurposing research methods, such as
video-occlusion, into training methods (Fadde, 2009) that increasingly bring
perceptual and decision skills into the realm of deliberate practice (Fadde &
Jalaeian, 2019).

One of the benefits of turning research methods into training methods
is that they tend to be highly targeted on specific skills. Typically, research
tasks tend to be as lean (i.e., inexpensive) as possible while validly
activating the target cognitive processes. Researchers need to repeat trials
reliably. They need to measure results precisely. However, most sports science laboratories can't afford to purchase or produce full-scale simulators.
So, led by economic necessity as well as scientific theory, sports expertise
researchers have developed methods to train critical components of expert
performance. In so doing, sports expertise researchers have created an efficient way to systematically train perceptual-cognitive and decision skills
that are often thought to come only from talent or vast experience.

However, with the commercial availability of virtual reality headsets and 360° cameras, more advanced technologies are being introduced into the laboratory setting to provide athletes with environments with high physical fidelity – meaning that it looks more real to the user.

Another source of training decision skill stems from flight simulators. Pilots stated that the simulator felt slow and unrealistic, therefore researchers incorporated a method called, Above Real-Time Training (ARTT), that incorporates faster training speeds (Guckenberger, Uliano, & Lane, 1993). The added time pressure creates a training environment with high psychological fidelity – meaning that it feels more real to the user. While ARTT's roots are in aviation, it has been proven to apply to other simulated contexts, including sport. Studies involving rugby players and baseball umpires have found improved decision-making skills compared to groups training at normal speed (Boatright, Kass, & Blalock, 2019; Lorains, Ball, & MacMahon, 2013).

While the focus has been on athletes, it is of note that perceptual-cognitive and decision training is applicable to sports officials as well. With officials having to make decisions on the biggest of stages (e.g., World Series, Super Bowl, NBA Finals) in real-time, it is remarkable that their calls are as accurate as they are. However, the good calls rarely receive notoriety, and the bad calls live in infamy. For a sport official, it is a stressful experience to make a wrong call and can lead to subsequent worse decision performance, which can be observed in the "home field advantage" effect (Nevill, Balmer, & Williams, 2002; Voight, 2009). By incorporating decision training, sport officials have an additional way to prepare outside of live competition.

In an era of emerging performance enhancement methods including virtual reality and biometrics, simple research and training methods developed in sports expertise research often provide the most scientifically validated approaches. On a theoretical level, the chess experiments that influenced expertise studies in general and sports expertise in particular drew upon Cognitive Information Processing (CIP) theory that views human cognition in terms of computer functions: input-processing-output. CIP provided a theoretical foundation for de-coupling the perception-action link in order to isolate, test, and train perceptual-cognitive and decision skills separate from psychomotor execution. CIP is not unanimously accepted as a model for sports expertise. Sports expertise researchers that adopt an ecological dynamics approach favor a direct perception model that postulates a direct link between perception and action, without intervening cognitive processes. In their view, that perception-action link cannot be broken without changing the essence of the skill.

The two theoretical stances have different implications for deliberate practice aimed at accelerating sports expertise. For instance, ecological dynamics aligns more with virtual reality simulation in which a perceptual-cognitive skill such as pitch recognition is linked with execution of the psychomotor skill during training. On the other hand, CIP aligns with part-task approaches such as video-occlusion in which pitch recognition is isolated for targeted training and then re-coupled with psychomotor batting skills during performance. Ironically, more immersive whole-task training is probably more appropriate for early skill acquisition while part-task perceptual-cognitive training is adequate, even optimal, for accelerating advanced athletes to the highest levels of sports expertise.

4.7 Benefits of Training Perceptual-Cognitive and Decision Skills

Of course, the main benefit of incorporating perceptual-cognitive and decision training is to expedite expertise, however, there are many other additional benefits working in their favor. For example, if an athlete can save wear-and-tear on their body by taking the same mental reps using portable devices (i.e., iPad, laptop, virtual reality headset) to train anywhere and without the physical action, that could be beneficial in extending their career.

While player safety is one of the most important benefits, many of these sports training technologies are *efficient* – in both time and cost. Now, multiple players and coaches are not required for an athlete to practice a specific skill (e.g., pitch recognition) and less time is required to master the skill. With wider accessibility to technology, the cost of entry to these technologies is reduced and could help lower income athletes or lower-level collegiate athletic programs gain an edge by complementing their experiences.

4.8 Future Research and Application of Perceptual and Decision Training

There remains a substantial, and unfortunate, research-to-practice gap in the theories, findings, and actual training methods of sports expertise research. Techniques such as video-occlusion are validated by decades of use in research laboratories, and a growing body of studies that demonstrate the effectiveness of perceptual and decision training. At this point, the ball is in the court of practitioners. As athletes, coaches, sport organization administrators, and – especially – intercollegiate athletics departments implement perceptual and decision training technologies they will be open to either conducting, or allowing researchers access to conduct, meaningful measures of intervention details and effectiveness. On the other hand, the sports expertise field needs to adjust publication expectations in order to encourage "fit in field" studies that may lack rigorous experimental control but that offer authentic training of competing athletes. When that happens, we expect that training techniques and technologies for perceptual and decision skills will rapidly be included in thousands of hours of deliberate practice, thereby accelerating the expertise of athletes across, and even beyond, sports.

References

Abernethy, B. (1988). Visual search in sport and ergonomics: Its relationship to selective attention and performer expertise. *Human Performance, 1*(4), 205–235. doi:10.1207/s15327043hup0104_1

Bloom, B. (1985). *Developing talent in young people.* New York: Ballantine Books.

Boatright, S. W., Kass, S. J., Blalock, L. D. (2019). *Improving the fidelity of training for baseball umpires* (Order No. 27663610). Available from Dissertations & Theses @ University of West Florida – FCLA; ProQuest Dissertations & Theses A & I; ProQuest Dissertations & Theses Global. (2352654850).

Chase, W. G., & Simon, H. A. (1973). Perception in chess. *Cognitive Psychology, 4*(1), 55–81.

Ericsson, K. A. (2003). The development of elite performance and deliberate practice: An update from the perspective of the expert performance approach. In Starkes, J., & Ericsson, K. A. (Eds.), *Expert Performance in Sports: Advances in Research on Sport Expertise,* (pp. 49–81). Champaign, IL: Human Kinetics.

Ericsson, K. A. (2006). The influence of experience and deliberate practice on the development of superior expert performance. *The Cambridge Handbook of Expertise and Expert Performance, 38,* 685–705.

Ericsson, K. A., Krampe, R. T., & Tesch-Römer, C. (1993). The role of deliberate practice in the acquisition of expert performance. *Psychological Review, 100*(3), 363–406. doi:10.1037/0033-295X.100.3.363

Ericsson, K. A., & Pool, R. (2016). *Peak: Secrets from the new science of expertise.* Boston, MA: Houghton Mifflin Harcourt.

Fadde, P. J. (2006). Interactive video training of perceptual decision-making in the sport of baseball. *Technology, Instruction, Cognition, and Learning, 7*(2), 171–197.

Fadde, P. J. (2009). Expertise-based training: Getting more learners over the bar in less time. *Technology, Instruction, Cognition, and Learning, 7*(2), 171–197.

Fadde, P. J., & Jalaeian, M. (2019). Learning with zeal: From deliberate practice to deliberate performance. In P. Ward, J. M. Schraagen, J. Gore, & E. M. Roth (Eds.), *The Oxford handbook of expertise* (pp. 927–950). Oxford, UK: Oxford University Press.

Fadde, P. J., & Zaichkowsky, L. (2019). Training perceptual-cognitive skills in sports using technology. *Journal of Sport Psychology in Action* (online). 10.1080/2152 0704.2018.1509162

Furley, P., Schweizer, G., & Bertrams, A. (2015). The two modes of an athlete: Dual-process theories in the field of sport. *International Review of Sport and Exercise Psychology, 8*(1), 106. 10.1080/1750984X.2015.1022203

Gladwell, M. (2008). *Outliers: The story of success.* New York: Little, Brown and Company.

Gray, R. (2019). Virtual environments and their role in developing perceptual-cognitive skills in sports. In A. M. Williams & R. C. Jackson (Eds.), *Anticipation and decision making in sport* (pp. 342–358). Oxford, UK: Routledge.

Guckenberger, D., Uliano, K., & Lane, N. (1993). *Teaching high performance skills using above real-time training.* (NASA Contractor Report 4528). Edwards, CA: Dryden Flight Research Facility.

Larkin, P., Mesagno, C., Spittle, M., & Berry, J. (2015). An evaluation of video-based training programs for perceptual–cognitive skill development: A systematic review of current sport-based knowledge. *International Journal of Sport Psychology*, *46*, 555–586.

Lorains, M., Ball, K., & MacMahon, C. (2013). An above real time training intervention for sport decision making. *Psychology of Sport and Exercise*, 14(5), pp. 670–674. doi:10.1016/j.psychsport.2013.05.005

Miller, G. A. (1956). The magical number seven, plus or minus two: Some limitations on our capacity for processing information. *Psychological Review*, *61*, 81–97.

Müller, S., Fitzgerald, C., & Brenton, J. (2020). Considerations for application of skill acquisition in sport: An example from tennis. *Journal of Expertise*, *3*(3), 175–182.

Nevill, A. M., Balmer, N. J., & Williams, A. M. (2002). The influence of crowd noise and experience upon refereeing decisions in football. Psychology of Sport and Exercise, *3*(4), 261–272. doi:10.1016/S1469-0292(01)00033-4

Roca, A., & Williams, A. M. (2016). Expertise and the interaction between different perceptual–cognitive skills: Implications for testing and training. *Frontiers in Psychology*, *7*, 792.

Voight, M. (2009). Sources of stress and coping strategies of US soccer officials. *Stress & Health: Journal of the International Society for the Investigation of Stress*, *25*(1), 91–101.

Williams, A. M. (2020). Perceptual–cognitive expertise and simultion-based training in sport. In N. J. Hodges & A. M. Williams (Eds.), *Skill acquisition in sport*, 3rd ed. (pp. 235–254). Oxford, UK: Routledge.

Williams, A. M., Ward, P., Smeeton, N. J., & Allen, D. (2004). Developing anticipation skills in tennis using on-court instruction: Perception versus perception and action. *Journal of Applied Sport Psychology*, *16*(4), 350–360.

Zaichkowsky, L., & Peterson, D. (2018). *The playmaker's advantage: How to raise your mental game to the next level.* New York: Simon & Schuster.

5 Collecting and Presenting Data for Performance Impact in Elite Sport: Finding the Right Balance between Humans and Machines

Elaine Tor, Emily Nicol, and Alice Sweeting

5.1 Introduction

The scientific analysis of sport performance is commonly referred to as performance analysis. The main aim of performance analysis is to advance the understanding of game or race behaviour in order to improve future outcomes (McGarry, 2009). Throughout the past ten years, as technology continues to progress, performance analysis has become commonplace amongst most professional team sports and Olympic Sports. More often than not performance analysis can provide coaches and athletes with a competitive advantage.

Elite sport has become a data rich environment with a large number of sport scientists working in the background to find the competitive edge that will assist athletes. In order to understand sporting environments and to make more objective decisions, heavy reliance is often placed on data collected from sophisticated performance analysis tools. However, large amounts of data can be detrimental to performance if not used in an effective manner. Data collection tools are now more accessible, cheaper and faster than ever before. Given the increased adoption of new technologies by sport organisations, sport scientists are now challenged to consider the different data types before analysing and interpreting this data. Added to this challenge is understanding the complexity of the sport environment and how to integrate both humans and machines in the decision making process (Robertson, 2020).

Data collection for performance analysis should assist coaches to make objective decisions about their athletes' behaviour. The selection of data is vital to increasing the impact it could potentially have on performance. Both qualitative and quantitative data collection is widely used in elite sport (Hughes & Bartlett, 2002). The type of data that is collected is largely contextual to the coach and athlete's needs. With such large amounts of data being collected in the sport environment, performance analysis is no longer just limited to game analysis (Hughes & Bartlett, 2002). Data can be collected for performance analysis in all disciplines of

DOI: 10.4324/9781003205111-5

sport science. This chapter will detail several important considerations to data collection methods when working in an elite sport setting.

Sport Science practitioners typically collect information around performance indicators to assess the performance of an individual, team or elements of a team (Hughes & Bartlett, 2002). Performance analysis has previously been reserved for team sports where athletes perform predominantly open skills. However, performance analysis can now be conducted on closed skills such as swimming, running, throwing and jumping. Further, these skills can now be assessed outside of a lab-based setting, providing more ecologically valid insights for sport scientists. Players can also be judged subjectively upon visual inspection, although these observations are best accompanied with objective data collection throughout the game/performance.

When selecting metrics of sports performance, how the data relates to an athlete's actions and outcomes needs to be understood at a scientific level. Glazier (2010) stated that there are many issues in sport today with scientists unable to adopt the appropriate theoretical framework on which to connect research and practice. This gap decreases the objective nature of data collection and can diminish the power of performance analysis. Sport Science practitioners must have an understanding of the mechanisms behind the data collected in all sporting environments in order to have an impact on overall performance. Impact can be measured in terms of performance improvement or injury prevention.

Data visualisation, in the 20th century, is considered a multi-discipline research area, complimented by a variety of tools for a diverse range of data types and methodologies (Friendly, 2008). In sport science, data visualisation is a tool to communicate analytical results and support decision-making processes by coaches, athletes and performance staff. Sport scientists now (more than ever before) have more access to large volumes of data, that can be quickly captured from differing technologies (Robertson, 2020). This data can be further characterised by volume or the amount of data generated, variety including the different types or sources and velocity or the speed of capture and stream of information (Kaisler, Armour, Espinosa, & Money, 2013).

Another attribute of data is complexity, or the measure of the degree of inter-relatedness whereby a small change may cause a ripple effect downstream (Kaisler et al., 2013). Sport, much-like the data captured in, is an environment rich in complexity (Robertson & Joyce, 2019). For example, consider a performance analyst is curious about athlete behaviours and interactions during a competitive netball match. This specific example would require capturing skilled actions or task constraints, likely via notational analysis. The collective behaviour of athletes could be captured via wearable technologies, such as local positioning systems. However, both data sources would need to be time-aligned together, analysed in an appropriate manner and presented in a meaningful way, in an attempt to delve deep into the complex problem. Therefore, the ability of the sport scientist to turn the ever-increasing volumes of data captured into usable pieces of information, by analysing and presenting data, might be considered a challenge and opportunity to upskill.

Performance analysis and data visualisation tools are now more accessible than ever. There are multiple applications that are readily available on smartphones and tablets. However, achieving sporting success through data collection is not as simple as it sounds. There are many factors that need to be given consideration in order to collect the right metrics, present and apply them in a meaningful way. Modern sophisticated data analysis tools allow sport scientists to create powerful visualisations. This book chapter will explore how data can be collected and presented to have an impact on performance in elite sport environments. Data analysis theories will be detailed with some practical elite sporting examples.

5.2 Question Framing

Question development is a key stage of the research process as it informs all subsequent research stages including methodology selection, data analysis and data visualisation or reporting (Bryman, 2007). Questions in performance analysis should be framed around performance indicators specific to the sport. A performance indicator is a selection or combination of variables that aims to define an aspect of performance (Hughes & Bartlett, 2002). Performance indicators or outcome measures can be used to assess an individual, team or elements of a team. They are broadly divided into two categories: technical and tactical. Technical measures quantify movement such as kinematics and kinetics. Tactical measures are parameters that describe match play or race pacing. Each of these categories can be used separately or together to convey important information surrounding an athletes' performance.

A good research question should be developed with consideration of several factors. These factors are listed below:

The current problem. The first stage to writing a research question is identifying an existing issue or problem. For example, a netball coach does not understand why their team consistently fails to convert a turnover. Or maybe a data analyst identifies that a large number of defensive football players fail to hit a leading target from the left back pocket. Some key questions to consider in order to better understand the context of a problem are:

- What is the topic area of the current problem?
- Who identified the problem?
- Why is the problem an issue?
- When does the problem occur?
- What could be improved or resolved from fixing the problem?

- What are the key performance indicators?
- What underlying theoretical framework currently exists?

The existing literature. An abundance of literature exists in the fields of sports science and data analytics. Exploring the existing literature enables the researcher to broaden their knowledge in the topic area and better understand how their current problem fits within the existing knowledge space. Several research problems may be completely or partially resolved by thoroughly searching and understanding the existing literature. Some key questions to consider in order to best utilise the existing research when formulating a research question:

- What literature currently exists in the area of the current problem?
- Has a previous research group investigated the same or a similar research problem? Can the problem be completely or partially resolved from the existing knowledge base?
- How is the existing literature similar or different to your current problem?

The intended audience. The intended audience plays a major role in question framing and development. The target audience determines the language used to frame the question, the breadth of understanding needed and the methodological rigour required to answer the current problem. When working within the sporting environment the intended audience must include the involved coach and/or athletes. Many research problems within the field of sports science directly relate to coach and/or athlete performance. Coach and athlete involvement is consequently required throughout the research process in order to better understand or resolve these issues. One of the greatest challenges faced by Sport Science Practitioners in the research context is attaining and maintaining coach/athlete buy-in. This may be aided through shared understanding between the researcher, coach and athlete regarding the importance of resolving the identified problem or issue. Some key questions to consider when framing the problem to coaches and athletes:

- How will resolution of the problem benefit the coach and athlete?
- How can the benefits of problem resolution be best communicated with the coach?
- Can some preliminary data be shared with the coach to support the problem's worth?

Once the current problem is well-understood, the existing literature has been consulted and the intended audience considered, the research question can be formulated. The research question(s) should be written clearly to ensure it is easily understood without further explanation. It should be phrased specifically to the identified problem, the population of interest and to the parameters of investigation. The question should also be worded in a way to elicit a complex response rather than a simple 'yes' or 'no' outcome. Instruction words such as *explore, examine* and *evaluate* are good examples of words that promote in-depth thinking and conversation.

The research question should inform all subsequent decisions made during the research process. Doing so will ensure the question is fully answered and conclusions drawn are of best use to the intended audience.

5.3 Data Collection

There are several methods currently available to collect data for performance analysis. Technology has allowed practitioners to progress from laboratory testing to in situ data collection. In situ data collection allows information to be far more ecologically valid and therefore can increase its impact on overall performance. Methods range from simple timing using a stopwatch to more sophisticated machine learning solutions. The constant need to collect and process data rapidly and at minimal cost has made selection of the ideal data collection tool paramount. The selection of data collection methods is highly dependent on a number of key factors.

Some key questions to consider when selecting a data collection method:

- *What is the research question?* Data collection methods must be selected with consideration of the key performance indicators, the specified performance context and phrasing used within the research question to ensure it can be answered adequately.
- *What resources are needed? Are these resources available?* Data collection method selection is partially dependent on resources available to the research team and/or the potential to obtain additional resources. Data collection resources may include equipment, personnel, research consumables, participants and an appropriate location.
- *Where will data collection occur? Will data collection occur in a lab, training or competition environment?* The data collection context is largely dependent on the requirements of the research question. In addition, some data collection methods are only available in specific environments. For example, wearable devices such as global positioning systems (GPS) or inertial measurement units (IMU) units are not permitted in several competition environments. The use of these data collection methods would consequently be inappropriate for the assessment of competition performance in this context.
- *Is information required instantaneously? Can additional time be taken to process the data?* Several data collection methods provide data in real-time for instant feedback provision to the coach and athlete. Other methods require an analyst or trained operator to process the data away from the data collection environment. The choice to utilise a data collection method that provides information in real-time or requires post-processing is dependent on a number of factors including the

context of data collection, purpose of data collection and the required accuracy of information.

- *What data collection methods have previously been used to collect similar information?* Data collection methods that have previously been used in the identified area of study may be found through consultation of the existing literature. The use of methods that have previously been used in the same area of study assists in the comparison of research findings between investigations. Consultation of the existing literature can also be used to establish the validity and reliability of some data collection methods.
- *Is the data collection method valid and reliable?* The data collection method must measure what it intends to measure and must provide repeatable results. This ensures a 'true' measure of each parameter is collected and results may be repeated at a later date. Information regarding methodology validity and reliability can often be found in various academic journals or pilot testing must be conducted prior to data collection in order to establish the validity and/or reliability of a method.

Sport Science practitioners can use a range of different data collection methods depending on the type of question that needs to be answered. Data collection methods can be classified into a number of categories which are as follows:

5.3.1 Qualitative vs Quantitative

Data collection typically falls into two broad categories. Qualitative data is typically subjective and description in nature, such as performance evaluations, interviews or technical observations. Quantitative data is information about quantities and numbers and is almost always objective, such as GPS data, computer vision, split times, kinematics and kinetics. In elite sport, it is commonplace to use a combination of both types of data collection depending on the situation, audience and intended message to be conveyed.

5.3.2 Notational vs Movement

There are also two main methods of data collection. Notational analysis is the process of counting predetermined events relating to performance such as goals, possessions and turnovers in team sport. Movement analysis is the process of tracking a players' movement around the sport field, GPS is a common tool to measure this. Television broadcast networks frequently utilise these types of data collection to increase fan engagement.

Notational and movement analysis is not just limited to team sports, these types of data collection methods are also widely used in individual sports. Sports such as tennis, Formula1 racing and swimming are just some examples of sport that rely heavily on data to assess performance.

5.3.3 Discrete vs Continuous

In relation to sports performance, data almost always falls into either discrete and continuous variables. Discrete variables are countable parameters such as game possessions, stroke or step counts. These are also sometimes referred to as aggregate measures. Continuous variables are measures taken over time or multiple measures taken over a given period. An example of continuous measures are heart rate traces over the duration of a game, distance travelled throughout a game or average velocity throughout a race section. Discrete and continuous data can also relate to once-off singular data collection or collection data over a number of different occasions for comparisons.

5.3.4 Data Collection Frequency

Another important consideration when collecting performance data is the frequency of data collection. Data collected from a single instance should be interpreted with caution as it may not be representative of the typical event. Data should instead be compared within an individual over multiple sessions or with other individuals in order to give more context to the performance. For example, global positioning system (GPS) data from an AFL half back on game day reports that the player covered a total of 14.2 km. From this data it could be concluded that the player averages 14.2 km per game. Over the next three weeks GPS data reports that the player covered 17.92 km, 16.7 km and 16.5 km each game. With consideration of multiple data collection points it is apparent that the initial average distance estimate underestimated the typical distance covered by the player. Instead, assessing the spread of the data could give more information on what a "typical" range is for the athlete.

5.4 Data Visualisation

Data visualisations are a medium to communicate and present data. The visual representation of information and data as figures or graphs are likely superior to tables, given the decreased cognitive load that is required to automatically process perceptual qualities (Kale, Nguyen, Kay, & Hullman, 2018). However, tables may be useful when the volume of data to be communicated is small or less than 20 observations. Similarly, tables may be useful if the objective of the data representation is to make comparisons between absolute values (Spence & Lewandowsky, 1991). Sport scientists should therefore be mindful of the intended objective of the visualisation or table, along with understanding the number and structure of underlying data observations. An understanding of data visualisation theory and upskilling in tools used to analyse and visualise data are essential for the modern sport scientist.

A number of tools exist to help sport scientists analyse and visualise data. Examining workflow or pipelines for sport science data analysis and visualisation is a good starting point in deciding which tools to utilise. For example, the typical data workflow of a performance analyst or sport scientist may include the capture of athlete tracking data, including GPS, via a bespoke program. The raw data is then (often, manually) exported into a spreadsheet program, whereby the sport scientist or performance analyst will usually perform basic data querying, including the inspection of outliers and statistical analysis. Data visualisation may also be performed in the spreadsheet program or the analysed data is exported to a specific program, such as GraphPad® or SigmaPlot®. However, sport scientists and performance analysts should be aware of the dangers and risks associated with data analysis (and subsequently, visualisation) in spreadsheet programs. For instance, upon an audit of 13 real-world spreadsheets, an average of 88% contained errors (Panko, 1998). Common issues with spreadsheets, which do not bode well for data analysis and subsequent visualisation, include inconsistent naming, extra white spaces between characters in cells, misrepresenting dates, incorrectly coding missing data, irregular data sheets and performing calculations on raw data files (Broman & Woo, 2018). If using spreadsheets as a multipurpose tool for data analysis and visualisation, sport scientists are encouraged to keep a copy of all raw data in plain text files, create multiple backups of data, be consistent in naming convention and use rectangular, or a series of rectangles, when storing data (Broman & Woo, 2018). For more tips and key principles on data organisation in spreadsheets, readers are encouraged to see Broman and Woo, 2018.

An alternate to performing data analysis and visualisation in spreadsheets, is open-source programming languages. These languages, including R, Python or Ruby, help the sport scientists to assess the underlying structure of their data, perform statistical analysis and create beautiful, informative data visualisations, dashboards and applications. Open-source data tools allow for flexibility, creativity and reproducibility in analysing and visualising sport data. Programming languages, including R (Team, 2013) and Python (Van Rossum, 1991), also enable the importing, analysis, modelling and visualisation of data in one workbench or platform. For sport scientists, a benefit of learning an open-source programming language is that data can be directly imported, analysed and visualised in the one program, rather than manually transforming the data across multiple platforms. This workflow also allows for reproducibility and enhanced collaboration, particularly when combined with version control software like Git and hosting services such as GitHub. Similar to attaching a file to an email or sharing a document with colleagues, GitHub works by committing versions of code, data and visualisations. Such practices allow for code, data and outputs to be revisited or reverted, by sport scientists or their colleagues, at a later date. For more on reproducibility and version control, readers are encouraged to see Bryan (2018). Another additional benefit of sport scientists learning or using a programming language is access to the large range of community created packages, specifically focused on data analysis and visualisation.

Data visualisation is a part of the analysis process, alongside tidying and modelling. At all stages of the process, sport scientists should attempt to better understand the data they capture. In R, the ggplot2 (Wickham, 2011; Wickham et al., 2019) package is designed for data visualisation and sits within the tidyverse (Wickham, 2011; Wickham et al., 2019), a collection of packages which use the same human-centred design for solving data science challenges. These tasks or challenges, which sport scientists often perform on a daily basis, typically include the import, tidying, manipulation and visualisation of data. For example, the abovementioned capture of GPS data requires importing into the R environment, inspection of outliers and tidying, perhaps according to athlete playing position. The data then needs to be visually communicated to coaches, support staff and athletes, potentially via a scatter plot. The abovementioned workflow can however all be completed in R, using the tidyverse (Wickham et al., 2019). The creation of visualisations, using R and packages such as ggplot2 as opposed to a spreadsheet program, also allows for reproducible and automated work. For example, using ggplot2, the sport scientist creates a scatterplot that visualises the total and sprint distance covered by ten team-sport athletes during a training drill. Once the visualisation is sent to a coach post-training, the same code used to create the graphic can also be deployed by the performance analyst, who adds a layer for the number of skilled involvements each player had during the training drill. This workflow allows for automation in creating the figure, rather than time spent clicking, dragging and dropping in a spreadsheet program whilst reducing the amount of duplicated code within the team.

Data visualisation R packages, including plotly (Sievert, 2020), RMarkdown (Allaire et al., 2020), flexdashboard (Iannone, Allaire, & Borges, 2018) and Shiny (Wickham et al., 2019), also allow for interactive reports, dashboards and custom-built applications to be deployed. This flexibility also allows for the sport scientist to not only customise visualisations but the format of the communicated data. For example, a coach may be interested in a static figure or report yet when presenting data to a group of performance support staff or athletes, an interactive visualisation or report may be more suitable for data presentation. Whilst there is a steep learning curve of a programming language and subsequent packages, including R and ggplot2, for data visualisation, the flexibility, reproducibility and creativity associated with these tools can be empowering for the modern sport scientist, who needs to capture, analyse and visualise data on a daily or weekly basis.

5.4.1 Considerations for Data Presentation

When presenting data to an individual or group of individuals, there are a number of considerations that sport scientists should be mindful of, including but not limited to:

- Considering what is the intended message of the visualisation?
- What is the intended context for use?
- What do people, who will read the visual, likely already know about the topic/ data?
- How will the data visualisation likely be used?
- How long does the visualisation need to be created in?
- What events, contextual factors or variables may impact the data collected?

As Edward Tufte (Tufte, 1985) once noted that *Excellence in statistical graphics consists of complex ideas communicated with clarity, precision, and efficiency,* sport scientists should also strive to create truthful, functional, beautiful, insightful and enlightening visualisations (Cairo, 2016). Whilst there are many books and journals, on theory and debate about what constitutes good design, readers are encouraged to consider the C.R.A.P (Contrast, Repetition, Alignment, Proximity) design principles by Williams (2015). The first of these principles, contrast, states thaat if two items are not exactly the same, they should be very different. For example, typographic families and weights, contrasting size and colour (Williams, 2015). Designing for accessibility and usability is also an important principle for sport scientists to incorporate. In sport, the colours red and green can often be used in data visualisations to represent a loss or win, bad or improved performance, respectively. However, given 8% of men and 0.4% of women are estimated to have some form of colour-vision deficiency from birth (Spalding, 1999), sport scientists should be careful in the selection of colours and palettes for visualisations. The Viridis (Garnier, Ross, Rudis, Sciaini, & Scherer, 2018) R package contains a variety of colour scales that are aesthetically pleasing, representative of data, easy to read by those who are colour-vision deficient and print well in grayscale.

Repetition is the second principle (Williams, 2015) which refers to repeating an aspect of the design throughout the entire project, report or visualisation. This includes colours, colour palettes, font families, weights and sizes, along with graphical elements and alignments. For a demonstration on how to share lines and alignments in figures, along with these C.R.A.P design principles, readers are encouraged to view the Data Visualisation presentation by Andrew Heiss (Heiss, 2020). The final design principle refers to proximity, whereby related items are grouped together and white space/ location are used to ensure groups are visually dissimilar. Together, these C.R.A.P principles provide sport scientists with a guideline on how to create clear, accessible and informative visualisations, using an easy to remember abbreviation!

The presentation of data is critical in elite sport, whereby athletes, coaches and performance staff use data to inform decision making on training or competition strategies. Summary statistics, such as the mean or median distance covered by a team-sport athlete during a training session, are easy to communicate, providing more detailed information about the distribution

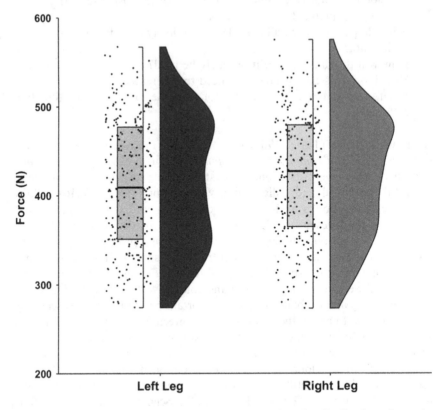

Figure 5.1 Example of a raincloud plot, demonstrating the distribution of an athlete's left and right leg Force (N), as collected over time.

of a dataset and including all data points. The presentation of raincloud plots (see Figure 5.1) is a neat way of visualising summary statistics and the underlying distribution, along with all data points collected (Weissgerber, Milic, Winham, & Garovic, 2015). Sport scientists should also remember that summary statistics are only useful when there are enough data to summarise and by displaying the full dataset. Readers can interpret any underlying distributions and statistical assumptions, which is especially important when working with continuous data (Weissgerber et al., 2015). Given the accessibility of open-source tools and packages, sport scientists should feel empowered and encouraged to be curious about the data they capture and analyse.

5.5 Applying Performance Analysis in Elite Sport

This chapter has explored many aspects of performance analysis. The diagram (Figure 5.2) summarises some of the key concepts and provides a

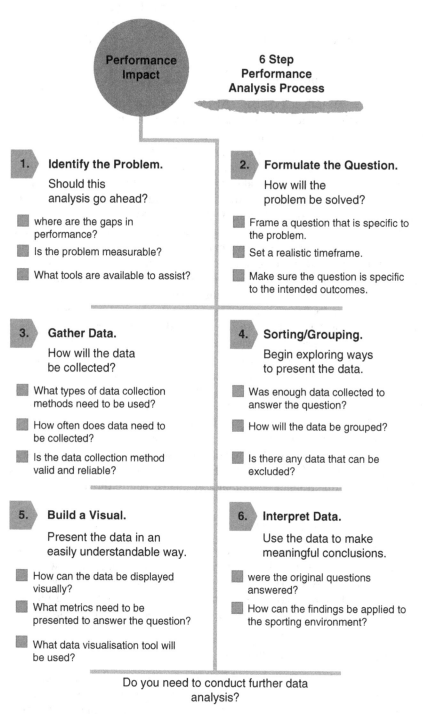

Figure 5.2 A visual summary of the performance analysis process outlined in this chapter.

practical way of conducting performance analysis. The key concepts can be applied to both team and individual sport settings. Two applied examples have been provided to demonstrate the concepts covered in this chapter.

5.6 Example – Team Sport

The design of Australian Football (AF) training sessions, to mimic and represent competitive matches, is a challenge faced by coaches and practitioners. This method of designing training is referred to as Representative Learning Design (RLD) and involves athletes sampling key constraints in training that are experienced in competition (Pinder, Davids, Renshaw, & Araújo, 2011). These constraints include information on an individual (e.g., an athlete's physical capacity), task (e.g., rules of a training drill) and the environment (e.g., the weather) categories (Davids, Button, & Bennett, 2008).

During AF training and competitive matches, an athlete's physical and skilled output can be captured. Information on an athlete's physical output, including distance and speed covered, can be captured via global positioning systems (GPS) or local positioning systems (LPS) whilst skilled output can be captured via notational analysis. Skilled output constraints can include the type (kick or handball), pressure (chase, frontal, physical or none), time in possession (less or greater than two seconds) and the effectiveness (effective or ineffective) of the disposal (Browne, Woods, Sweeting, & Robertson, 2020). The sport scientist or performance analyst typically needs to combine both data sources (physical and skilled) together to give an understanding of AF training and competition. This involves working with continuous (GPS or physical) and discrete (notational analysis or skilled) data sources to understand task constraints during training and competition.

A specific (hypothetical) example of how data can be collected in training and transformed into a useful tool for planning subsequent the training session is discussed in the following sections. In this example, kick effectiveness was collected during one training session and evaluated before the next.

5.6.1 Step 1. Identify the Problem

The head coach is interested in how new recruits are going in training. Specifically, they are interested in what their effectiveness is, for each new recruit, across a training session. The coach is also interested in how this kick effectiveness relates to other players of a similar age, playing position and physical capacity, along with how they compare relative to the entire playing list.

5.6.2 Step 2. Formulate the Question

By examining relevant literature, kick effectiveness could be impacted by performer, environmental and task constraints. Specifically, as all training sessions are held outdoors, the amount of rainfall preceding the training session, along with average wind speed and direction during each training drill, within the session of interest. Task constraints that could impact kick effectiveness include the drill type, for example, a small-sided game or match simulation. Performer constraints included the age of the athlete, for example, if they were under or over 20 years of age. Given these considerations, the following question was developed: How do constraints interact and influence kick effectiveness during AF training?

5.6.3 Step 3. Gather the Data

Video footage of (n = 10) AF athletes, from the same club, was collected during a single training session. This footage was subsequently analysed by performance analysts who coded all footage using a code window and video analysis software. Each kick disposal, for all players, was coded during training and rated either effective or ineffective, as per Browne et al. (2020). All data was subsequently imported into the R computing environment for analysis and visualisation.

5.6.4 Step 4. Sorting/Grouping

Individual athlete data was further coded based on each athlete's age (under or over 20 years of age) and playing position (forward, midfield, ruck or defender). Each drill was also coded (small-sided game or match simulation) and the average wind speed and direction recorded for each drill.

5.6.5 Step 5. Build a Visual

A summary table was first calculated, containing the average and total number of disposals and effectiveness for the session (see Table 5.1). However, given the purpose of the problem was to inspect how specific athletes compare to one another and the group, a scatterplot (see Figure 5.3) was subsequently created

Table 5.1 Athlete Disposal Count during Training and % Effective Disposals.

Athlete Name	Age (years)	Total Disposals	% Effectiveness
Alice	19	16	30%
Elaine	23	50	85%
Emily	18	20	33%
Michele	26	35	78%
Molly	24	47	82%
Ivy	19	19	47%
Ava	21	36	61%
Marni	25	42	79%
Sharyn	22	31	78%
Fay	24	40	81%

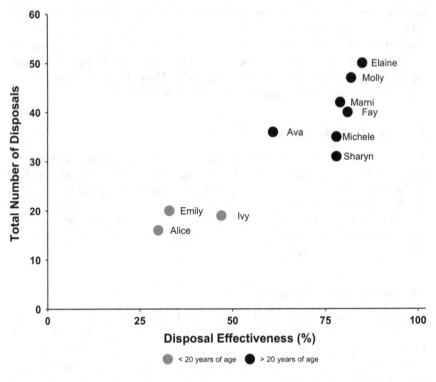

Figure 5.3 Athlete disposal count during training and % effective disposals, with colour displaying the category of athlete.

to visually represent the average and total number of disposals and effectiveness for the session. Using the ggplot2 and plotly packages, this figure also allows for added interactivity.

5.6.6 Step 6. Interpret the Data

Individual data of younger (less than 20 years of age) can be compared to each other and the wider (over 20 years of age) group. Based on this data, younger athletes may have less disposals during training, compared to older athletes. This information could subsequently reported to coaching staff, to ensure that younger players are gaining disposals during training.

5.6.7 Further Analysis

Constraint interactions could be modelled via association rules, to understand the impact of different constraints on disposal effectiveness. Further analysis could also explore the temporal nature and interaction of constraints on AF disposal effectiveness and type, whereby coaches and practitioners can examine the degree of representativeness between specific training drills, sessions and competitive matches. Finally, individual drills and disposal type could easily be added into the Table 5.1 and Figure 5.3, for a more complete analysis.

5.7 Example – Individual Sport

Swimming is an individual sport that relies heavily on performance analysis to inform training interventions and race strategies. During training, wearable devices can be worn to collect live heart rate and stroke data whilst in competition machine learning can be used to conduct race analysis. During competition, an infinite number of parameters can be calculated with each race. This information is often collated with race footage and transmitted wirelessly via cloud computing to the coaches' iPad. It is the role of the performance analyst and biomechanist to filter the data and provide the coach and athlete with the most important information. Typically, only parameters with immediate performance impact will be presented during competition whilst a more comprehensive analysis of the data will be presented following competition and used to influence future training interventions.

A specific example of how data can be collected in competition and transformed into a useful tool for training is dealt in the follwing sections. In this example, stroke timing data was collected from several breaststroke race videos.

5.7.1 Step 1. Identify the Problem

The team biomechanist identified several technical errors in their athlete's breaststroke technique. These errors were causing the athlete to decelerate considerably during various phases of the stroke. By fixing these errors it was thought that the athlete would be able to maintain a higher average velocity during competition.

5.7.2 Step 2. Formulate the Question

Through consultation of the relevant literature it was found that temporal pattern inefficiencies could be a cause of the observed technical error. Temporal patterns had been investigated by several research groups using simulated race pace efforts, however these had not been investigated during competition. With consideration of the identified problem and the existing literature, the following research question was developed: How do elite breaststroke swimmers manipulate temporal patterns to increase average velocity?

5.7.3 Step 3. Gather the Data

Methods used in this example were based on those previously described within the literature. Video footage from over 50 athletes across 10 elite swimming competitions was collected. This footage was subsequently analysed using video analysis software and used to identify the frame number at which specific stroke events occurred. Identified frame rates were then entered into a customised R Shiny application for the calculation of temporal parameters.

5.7.4 Step 4. Sorting/Grouping

Data was separated into four subgroups based on sex and race distance: 100 m female, 200 m female, 100 m male and 200 m male. Subgroup classification allowed temporal data from the individual athlete to be compared to athletes of the same sex and across the same race distance.

5.7.5 Step 5. Build a Visual

A bar chart was used to display the temporal data to coaches and athletes in an easy-to-understand format. Presentation of the data in this way enabled the coach and athlete to visually inspect the temporal information and understand how the stroke phases were coordinated. Applications like RShiny allow sport practitioners to create dynamic reports that provide options for coaches to either turn on or off certain parameters in order to make the data more user friendly (Figure 5.4).

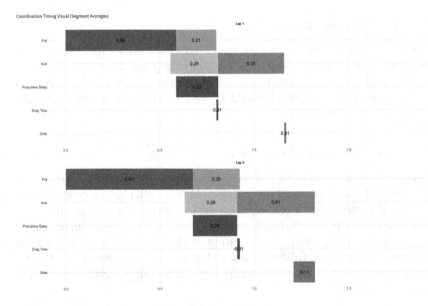

Figure 5.4 Example breaststroke timing analysis plot.

5.7.6 Step 6. Interpret the Data

Temporal data for the individual was compared to others within the same subgroup to identify phase differences. It was observed that the individual athlete spent a longer amount of time in a phase associated with rapid deceleration than other athletes within the subgroup. This information was subsequently reported to the coach and used in the development of a training program aimed at reducing time spent in this deceleration phase.

5.7.7 Further Analysis

Statistical analysis was conducted on the full dataset in order to identify trends that will aid further understanding breaststroke timing in elite swimmers.

5.8 Conclusion

For data to assist with performance enhancement, sport scientists must continue to find new ways of bridging the gap between data collection and data interpretation. Large amounts of data are commonly collected in most team and individual sporting environments, however the impact that data has on performance hinges on how the data is analysed, presented and interpreted. This chapter has explored a framework for data use in elite sport settings and detailed a number of data visualisation theories. In summary, collecting endless amounts of data without thorough understanding of the mechanisms behind the information presented will not be impactful on performance or even detrimental in some cases. Sport scientists must ensure they find the right balance between data collection and interpretation to ensure specific questions are answered surrounding an athletes' performance.

References

Allaire, J., Xie, Y., McPherson, J., Luraschi, J., Ushey, K., Atkins, A., ... & Iannone, R. (2020). Rmarkdown: Dynamic documents for R. *R Package Version, 2.*

Broman, K. W., & Woo, K. H. (2018). Data organization in spreadsheets. *The American Statistician, 72*(1), 2–10.

Browne, P. R., Woods, C. T., Sweeting, A. J., & Robertson, S. (2020). Applications of a working framework for the measurement of representative learning design in Australian football. *PloS One, 15*(11), e0242336.

Bryan, J. (2018). Excuse me, do you have a moment to talk about version control? *The American Statistician, 72*(1), 20–27.

Bryman, A. (2007). The research question in social research: What is its role? *International Journal of Social Research Methodology, 10*(1), 5–20.

Cairo, A. (2016). *The truthful art: Data, charts, and maps for communication.* New Riders.

Davids, K., Button, C., & Bennett, S. (2008). *Dynamics of skill acquisition: A constraints-led approach.* Human Kinetics.

Friendly, M. (2008). A brief history of data visualization. In *Handbook of data visualization* (pp. 15–56). Springer.

Garnier, S., Ross, N., Rudis, B., Sciaini, M., & Scherer, C. (2018). Viridis: Default color maps from 'matplotlib'. *R Package Version 0.5*, *1*, 2018.

Glazier, P. S. (2010). Game, set and match? Substantive issues and future directions in performance analysis. *Sports Medicine*, *40*(8), 625–634.

Heiss, A. (2020). Graphic design. Retrieved from https://datavizm20.classes. andrewheiss.com/content/02-content/

Hughes, M. D., & Bartlett, R. M. (2002). The use of performance indicators in performance analysis. *Journal of Sports Sciences*, *20*(10), 739–754.

Iannone, R., Allaire, J., & Borges, B. (2018). Flexdashboard: R markdown format for flexible dashboards. *R Package Version 0.5*, *1*.

Kaisler, S., Armour, F., Espinosa, J. A., & Money, W. (2013). *Big data: Issues and challenges moving forward.* Paper presented at the 2013 46th Hawaii international conference on system sciences.

Kale, A., Nguyen, F., Kay, M., & Hullman, J. (2018). Hypothetical outcome plots help untrained observers judge trends in ambiguous data. *IEEE Transactions on Visualization and Computer Graphics*, *25*(1), 892–902.

McGarry, T. (2009). Applied and theoretical perspectives of performance analysis in sport: Scientific issues and challenges. *International Journal of Performance Analysis in Sport*, *9*(1), 128–140.

Panko, R. R. (1998). What we know about spreadsheet errors. *Journal of Organizational and End User Computing (JOEUC)*, *10*(2), 15–21.

Pinder, R. A., Davids, K., Renshaw, I., & Araújo, D. (2011). Representative learning design and functionality of research and practice in sport. *Journal of Sport and Exercise Psychology*, *33*(1), 146–155.

Robertson, S. (2020). Man & machine: Adaptive tools for the contemporary performance analyst. *Journal of Sports Sciences*, *1*, 1–9.

Robertson, S., & Joyce, D. (2019). Bounded rationality revisited: Making sense of complexity in applied sport science. SportRxiv.

Sievert, C. (2020). *Interactive web-based data visualization with R, plotly, and shiny.* CRC Press.

Spalding, J. (1999). Colour vision deficiency in the medical profession. *British Journal of General Practice*, *49*(443), 469–475.

Spence, I., & Lewandowsky, S. (1991). Displaying proportions and percentages. *Applied Cognitive Psychology*, *5*(1), 61–77.

Team, R. C. (2013). R: A language and environment for statistical computing.

Tufte, E. R. (1985). The visual display of quantitative information. *The Journal for Healthcare Quality (JHQ)*, *7*(3), 15.

Van Rossum, G. (1991). Python. In.

Weissgerber, T. L., Milic, N. M., Winham, S. J., & Garovic, V. D. (2015). Beyond bar and line graphs: Time for a new data presentation paradigm. *PLoS Biol*, *13*(4), e1002128.

Wickham, H. (2011). Ggplot2. *Wiley Interdisciplinary Reviews: Computational Statistics*, *3*(2), 180–185.

Wickham, H., Averick, M., Bryan, J., Chang, W., McGowan, L. D. A., François, R., ... & Hester, J. (2019). Welcome to the tidyverse. *Journal of Open Source Software*, *4*(43), 1686.

Williams, R. (2015). *The non-designer's design book: Design and typographic principles for the visual novice.* Pearson Education.

6 The Design of Interactive Real-Time Audio Feedback Systems for Application in Sports

Nina Schaffert and Sebastian Schlüter

6.1 Introduction

In competitive sports and increasingly in amateur sports as well as in rehabilitation, testing and training devices are used to monitor and optimize the training process and to analyze and improve the athletes' performance (Liebermann et al., 2006; Brock & Ohgi, 2016). In elite sports, the evaluation of movement has become unthinkable without the use of systems for data measurement and analysis (James et al., 2006). These technologies offer coaches and athletes innovative and effective support. The presentation of objective data via technical equipment quantifies the information and enables the detection of deviations in an actually performed movement. With the aim of "perfect" or targeted movement execution from a biomechanically practical perspective, the subjective perception and the athlete's individual internal representation of the movement should always be considered. To fulfill the needs of a specific field, the measurement and analysis systems are developed by engineers in cooperation with scientists, biomechanics and coaches. The analyses provide the basis for the training concept and for the decision on the priorities of the training content.

Recently, the role of interactive augmented feedback has proven benefits of improving motor skills and movement techniques (Magill, 2011; Schmidt & Lee, 2019). Augmented feedback is defined as any additional information provided to the athlete beyond what the athlete can obtain intrinsically from the performance itself (Schmidt & Wrisberg, 2000). Suitable feedback systems can significantly contribute to shortening the acquisition time following the principle of objectively supplementation of fast and direct information.

Sports science research provides empirical evidence that the use of synthesized additional audio information (non-verbal representation of data in sound or data-driven sonification; Hermann, 2020) as interactive feedback has positive effects on the accuracy of perception, the reproduction and regulation of movement patterns as well as on motor control and learning (Schaffert, Janzen, Mattes & Thaut, 2019 for an overview). Here, the interface between human and computer becomes an integral part of the process of selecting, manipulating, or controlling the display. Termed as auditory displays or interfaces, such systems

DOI: 10.4324/9781003205111-6

have many attractive features. Notably the most important for practical ease-of-use: sounds essentially require no training, despite of the more complex uses of sounds discussed in sonification research (i.e., Ferguson, Martens & Cabrera, 2011; McGookin & Brewster, 2011), but simple sounds can be understood without training. Also, auditory displays are effectively accessible to most people (excluding those with a hearing impairment).

Following the current trend of quantitative analysis of human movements in various application contexts and the technological development in recent years, the use of inertial measurement units (IMUs) provide a suitable basis for mobile solutions of real-time measurement systems for in-field sports and rehabilitation performance analysis (Brock & Ohgi, 2016). The recent advances in miniaturized wearable technologies enables practical applications by reducing both the size and the power consumption of the devices as well as the limitations of other methods such as complex motion capture systems. Research in sports science has provided evidence for the use of IMUs to monitor human movements (James, Moritz & Haake, 2006). This technology allows for different set-ups and positioning on the athlete's body or on the sports equipment and enables reliable measurements in various environments, such as in water. In addition, other systems can be used to improve data acquisition (e.g., additional use of GPS data for location information).

This chapter aims at describing the design of interactive real-time audio feedback systems for sport applications. Our focus is on the development of specially developed sonification algorithms for real-time sound synthesis and the information transfer to the athlete with the presentation of practical applications in rowing and swimming

6.2 Background

The research field of sonic interaction design explores methods to convey information (including aesthetic and emotional qualities) using sounds in interactive contexts (Rocchesso & Serafin, 2009; Hermann, Hunt & Neuhoff, 2011). This research field deals with the challenges of creating sound-mediated interactions by designing and implementing novel interfaces for creating sounds in response to human gestures. Sonic interaction design is closely related to the subtopic of the human-computer interaction (HCI) field known as sonification (Hermann, 2008). Such a research field addresses how information can be conveyed in an auditory, usually non-speech, form. Basically, data from various sources are transformed into sound to better assist the listener in understanding and interpreting the data. Interactive sonification is then the study of human interaction with a system that converts motion data into sound, and is a subfield of sonification (Hermann et al., 2011). Here, the action-perception loop resulting from the interaction with the developed interfaces is investigated. The user discovers how his/her gestures and movements modulate the sound. In the so-called movement sonification, which is also part of interactive sonification, the challenge is to control the execution of movements in response to sounds. Here, the auditory feedback can guide the user's actions by providing information on how the user can modify the actions themselves. In contrast to graphical user interfaces (GUIs) that are traditionally used in sports technique training such as graphs or video, sonification is particularly attractive because communicating through sound does not interfere with an athlete's vision, maintaining required visual stimuli for performance.

Interactive real-time audio feedback within a movement context was proven beneficial particularly in cyclic sports (Schaffert, Janzen, Mattes & Thaut, 2019). Because the auditory system has a remarkable temporal resolution and sensitivity to frequency and time changes, it is possible to provide biomechanical feedback on kinematic parameters of the movement execution by changes in pitch and rhythm using algorithmically-defined sound sequences. Since one of the key points in simultaneous feedback is that the information content is relevant, accurate, fast, and easy to understand (Anderson, 2010), sonification must provide intuitively understandable auditory information to support the process of regulating action execution and motor learning (Wulf & Prinz, 2001; Lahav, Saltzman & Schlaug, 2007; Wulf, 2007).

The role of attention in motor control has been widely studied. The results suggest that there is a close relationship between attention and motor control to coordinate actions (Wulf & Su, 2007). Device-based feedback can direct an individual's attention to specific sections and facilitate the process of perception-action coupling (Lahav et al., 2007). The resulting sound information can alert athletes to their technique and allow them to focus on the correct execution of movement. In addition, with sound information, motor behavior can also be changed unconsciously (Bigand, Lalitte & Tillmann, 2008), which also makes it a promising tool for implicit learning.

6.3 Applications in Rowing and Swimming

6.3.1 Considerations and Requirements

According to Anderson (2010), feedback technology in biomechanical applications must be (i) accurate and relevant for the specific task, (ii) appropriately timed and delivered and (iii) identifiable by the athlete. To provide meaningful acoustic information, it is important to consider practical aspects with regard to the device's usability during training.

For the analysis of movements, particularly in high performance sports, the accuracy of the system is one of the most important factors in order to provide adequate feedback to the athletes about their movement execution for both, the post-session analysis but also for the real-time feedback situation. In the context of human movement the timeliness of the data transmission is important: a system with a considerable latency will be evaluated as unusable in any real-time multimedia application. In high performance sports, bio-mechanically driven feedback must be real-time to be useful and effective. Here, the challenge is to develop algorithmic solutions to synthesize motion data into sounds suitable for direct parametric control during motion execution (Fontana et al., 2011), as it is known that temporal delays of 150–180 ms of sound as a result of movement execution lead to systematic changes in the kinematic aspects of the movement (Menzer et al., 2010; Kennel et al. (2015). Thus, the sonification of motions and transmission of feedback information should take place with low latency, ideally < 30 ms).

A particularly important and sensitive aspect in high-performance sports with elite athletes is the wearability of the system: it should be unobtrusive and ideally small and lightweight to not interfere with the execution of movement.

In aquatic sports such as rowing and swimming, the devices require to be waterproof. The system should provide feedback to both athletes and coaches simultaneously and separately. This allows the coach to compare if and how the execution of the movement has changed during the sections with and without acoustic feedback. Therefore, it must be possible to remotely control the transmission of the sound in agreement with the coaches' concept.

From a biomechanical point with regard to the analysis of the training, the information presentation requires to demonstrate the content during training in different conditions. i.e., differences between individual move-ment cycles as well as between different cycle frequencies/rates should be audibly differentiable so that the changes are distinguishable in the sound produced. Also has the sound to be audible regardless of all surrounding acoustic events, such as naturally occurring sounds of the water, the boat or passing motorboats or other swimmers. At the same time the sound should not be obtrusive, disturbing or annoying, but comfortable to be heard.

It is important to model/interpret the sound according to the effects of changes in kinematic parameters (e.g., boat acceleration) in a predictable way, in so far as the sound result should be understandable and refer to naturally occurring sounds that are familiar to the athletes.

With regard to aesthetic aspects of sonifications, however, there is still a clear need for further development. Aesthetics in the sense of "cosmetics", as the common understanding of the term in auditory display, is in contrast to the functionality of the information content. It is well known that a pure tone presentation is difficult to listen to over a longer time period (Schaffert, Mattes, Barrass & Effenberg, 2009; Schaffert & Gehret, 2013). Here, the challenge is to balance meaningful and data-based information transformation without artificial data manipulations for the benefit of a more melodic sound result versus the design of a more pleasant sound sensation but with the dis-advantage of losing relevant information about the actual movement execution.

The following section describes the design and development of two acoustic feedback systems for rowing (Sofirow) and swimming (Sofiswim) and their use in training practice.

6.3.2 Rowing: Sofirow

6.3.2.1 Objective and Approach

Characterization of the Boat Motion

Maximizing average boat speed is the explicit goal of coaches and athletes, since the fastest team is undoubtedly the winning team in rowing regattas. From a biomechanical perspective, it is a clear mechanical factor to optimize rowing performance and technique (Baudouin & Hawkins, 2002; Kleshnev, 2020).

Distinctions are made for the boat movement. Besides the translational movement in the direction of travel (acceleration), which is the result of all forces acting on the system (boat and rowing equipment), there is the rotational movement of the boat around its fixed three axes: pitch ("tamping", rotation around the transverse axis), which characterizes the intra-cyclic lowering and raising of bow and stern, roll (rotation around the longitudinal axis), which characterizes the intra-cyclic tilt (front-back or bow-star), and yaw (rotation around the depth side/vertical axis of the boat), which characterizes the course deviation (Kleshnev, 2012).

All directions in common involve a change in the amount of the water-covered surface of the boat and thus, the water resistance, whereby fluctuations in boat speed and overall resistance result. Influencing factors on the boat motion are external conditions such as boat construction, wind and waves and also the athletes' movements and their execution (Formaggia, Mola, Parolini & Pischiutta, 2010). While the former must be taken as given, it is possible to influence the latter by improving the quality of the athletes' movement execution (athletes' mass displacement relative to the boat and dependent on the vertical foot-stretcher and blade forces, force-action coupling). Consequently, athletes are supposed to execute the several parts of the rowing movement (drive, recovery phase and their sub-phases) as consistently and smoothly as possible. Figure 6.1 shows the division of the rowing stroke cycle and the phase structure of the athletes' movements on the basis of the boat acceleration time curve.

Since an application for high-performance rowing using acoustic feedback has shown beneficial effects for the execution of the recovery phase (Schaffert, Mattes & Effenberg, 2011), it was assumed that an acoustic representation of the boats' rotational movement can contribute to reductions of intra-cyclic fluctuations in boat speed. The focus here is on the reduction of amplitudes which are directly represented in the amount and curves of the respective direction. It aims to support the athletes' feeling for the boat run and execution of the rowing motion by sensitizing their awareness for propulsive critical periods in the rowing cycle.

To obtain information on the boat motion in all its directions, and to develop approaches for its modification in training, regular data recording

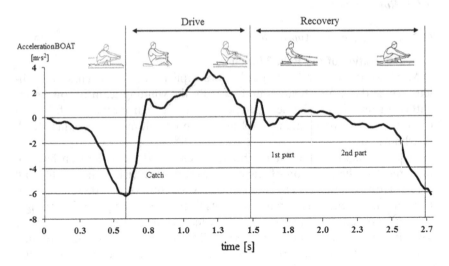

Figure 6.1 The rowing stroke with its two characteristic main phases: Drive and recovery (or release), which are further subdivided into the front and back reversal as the catch and finish turning points.

and analysis is necessary. A prerequisite therefore is a practicable measuring system that can be used in the racing boat and in daily training.

6.3.2.2 System and Sound Design

The device measures and stores propulsive boat acceleration and the boat's rotational 3D-movement (pitch, roll and yaw) with an IMU (sampling rate: 100 Hz). Measuring accuracy for acceleration is 1% (within ±2 g), measuring accuracy for rotation: 0.1°. In addition, boat speed is measured with a 4-Hz-GPS-sensor. The system calibrates itself in a rest position after the main switch was turned on. Two more switches control the measurement and sonification turn on/off, and a third toggle switch allows changing between the sonified parameters (acceleration, pitch, roll or yaw). Sound transmission was via a mini Coxbox (Nielsen Kellermann) that was used as an amplifier unit and to switch the sound on/off. Further, WLAN transmission also allowed online adjustments independent of the athletes if necessary. The athletes heard the sound over waterproof loudspeakers mounted inside the boat. Depending on the size of the boat, one or two speakers were needed to hear the sound despite the natural ambient noise. Figure 6.2 provides an overview of the system set-up.

A custom-designed sonification toolkit was used to map the data to sound parameters (sonification). This toolkit allowed an acoustic transformation of the boat motion data whereby the following general design criteria for the sonification algorithm were:

Figure 6.2 Overview of the Sofirow system: GPS sensor, power supply, amplifier unit and speaker placed on/in the boat.

1. Sounds synthesized in real-time in response to boat movement: the latency between the event that caused the sound and its presentation must be short enough for the athlete to perceive the information as coinciding exactly with the event. Real-time auditory feedback can be defined as the perceptual synchrony of a given movement and its auditory representation. The threshold for intermodal detection of asynchrony should ideally not exceed 100 ms (Van Vugt & Tillmann, 2014).
2. It should portray an accurate equivalent of the signal acquired: the feedback must provide an intuitive measure of the boat's motion so that the athlete can recognize the difference between the movement directions and between low and high levels of executed rowing cycles.
3. The feedback requires to be adjustable for the directions of the boat motion, and the data should be audible when rowing is performed at different speeds of motion.
4. The feedback must be simple enough that no learning is required. And at the same time it should be pleasant to listen to (or at least not annoying).

The final choice of sonification method was the modulation of the frequency of a square wave oscillator. Each data point of the movement parameter (acceleration, pitch, roll and yaw) was mapped to notes on the chromatic scale, with the intention of making the sound functional and at least somewhat pleasing (more instrumental than noise-like). This method also allows the transformation of data into an easy understandable resultant sound.

6.3.2.3 Testing

The results of several case studies in high-performance rowing with the German National Rowing Team (Juniors and Seniors as well as with the adaptive National Rowing Team) repeatedly found that movement sonification led to faster boat speeds, increased distances traveled per stroke, and improved crew synchronization compared to training without additional auditory information. This occurred, as hypothesized, as a result of fewer occurring fluctuations within the boat acceleration-time curve, as revealed by the intra-cyclic analysis. More specifically, the mean acceleration curves showed that the acoustic feedback influenced the temporal structure of the rowing cycle with respect to the recovery phase, especially its second part, as well as the front reversal (Schaffert, Mattes & Effenberg, 2011; Schaffert & Mattes, 2014, 2015, 2016).

The results of requests with the athletes showed an overall high acceptance of the acoustic feedback among athletes and coaches. The way of presentation was intuitively understandable and the sonification was rated as a functional training aid which becomes manifest in the individual statements: focused improvement of the weak points in the movement and smoother movement with the sound. Preferred presentation of the acoustic feedback was via speakers in order to be able to hear the natural soundscape without being isolated from sounds athletes were used to hearing while rowing.

Taken together, the sound enhanced the athletes' perception for the rowing movement in general and specific sections in particular. The medium of presentation, independent of vision, induced a feeling for the duration of the movement as well as it enhanced the perception of an optimally executed movement as the desired outcome.

The resulting sensitivity to time-critical structures in the rowing cycle yielded an improved synchronization of the crew as well as of each athlete's individually executed rowing technique, as had been assumed. This was confirmed on the one hand by the fact that the improvement was detected in all boats (including the single scull), and on the other hand, by the athletes' individual statements.

Characteristic phases of the boat's acceleration-time trace were perceived in the sound sequence as changes in tone pitch within the single rowing strokes as well as between the stroke series. And athletes' focus of attention was guided reliably by the sound to specific sections within the rowing cycle, particularly to the second part of the recovery phase and to the finish turning points. The results showed fewer intra-cyclic interruptions between the stroke series as well as within the single rowing stroke when rowing with the sound.

6.3.3 Swimming: Sofiswim

6.3.3.1 Objective and Approach

This section describes initial considerations for an application of interactive sonification for everyday use in swimming training with a focus on frequency and forward acceleration produced by the swimmer when performing the underwater dolphin kick (UDK). The aim is to represent the characteristics of the UDK in the sound provided by the real-time audio feedback system Sofiswim to enable congruence between sound, action, and reaction. The focus is on describing the design, followed by a brief summary of the evaluation and practical effectiveness, which was tested in a pilot study.

6.3.3.2 Characterization of the UDK

The UDK is a cyclic motion that is divided into downward (downbeat) and upward kick (upbeat). The downward motion of the feet is associated with an upward motion of the hip and vice versa for the upward kick. The downbeat starts when the feet reach their highest point and ends at the lower point. It is characterized by bent knees and feet aligned with calves. Forward propulsion is generated by deflecting the water from the feet backwards. The flexibility of the ankles is an important factor in propelling efficiency (Maglischo, 1993; Colman, Persyn & Ungerechts, 1998).

This is where the upbeat starts. The upbeat is initiated by a rebound-like action that pushes the thighs upwards. The legs are stretched in the first part of the upbeat and the lower legs and feet are relaxed.

Elite swimmers try to spend as much time as possible underwater performing the UDK to propel forward. During this phase, drag is reduced compared to swimming on the water surface. Novais et al. (2012) found that the highest efficiency to propel forward can be reached at a depth of 0.6 to 0.7 m. Thus, the ability to maintain maximum velocity during underwater kick is essential in elite swimming and one of the determining factors for overall performance (Cossor & Mason, 2001).

Theoretical considerations for the auditory display refer to the acoustic representation of the forward acceleration during the UDK. It must provide information about the propulsion generated by each kick to provide relevant information for its execution. More precisely, the information is most meaningful when changes in the sound reflect changes in the acceleration data that correspond to athletes' experiential expectations. At the same time, the consistency of the athlete's underwater movements during the UDK is an important factor for swimming speed and should be reflected in the sound information. To address both, we have decided to sonify the forward acceleration parameter during the UDK with two approaches: (1) continuous and (2) discrete sonification.

The forward acceleration parameter was chosen as the main factor to increase the swimming speed and thus improve the athlete's performance. Experienced swimmers know how to perform the UDK in general. However, some aspects of movement are difficult to perform and may attract the athletes' attention through the sound (e.g., vertical hip acceleration). The UDK movement may have some variations that result in better or worse forward acceleration. Common mistakes that negatively influence forward acceleration are: (1) a low hip and ankle flexibility; and (2) maintaining a narrow streamline position, both of which should result in a lower pitch of the tone produced, valid for both continuous sonification and discrete sonification; (3) a start-stop action should result in a non-rhythmic sound and a (4) decrease in kick frequency should result in a longer pause between each sound in case of the discrete sonification. In the continuous sonification version, the aim of (3) is to create a shorter duration between highest and lowest pitches, while (4) should lead to a longer duration between the sounds with the highest pitches, which is easier for the athlete to recognize than the lowest pitches.

By providing two alternatives of the sonified UDK, athletes can choose which benefits are best suited to their individual way of executing the UDK. Individual differences (e.g., anatomical or perceptual) are also taken into account, as each athlete has their own way of maximizing propulsion, regardless of an "ideal" UDK technique. Here, instead of a technical model, it was intended to provide information about the actual execution of the movement.

6.3.3.3 System and Sound Design

The analysis and audio feedback system "Sofiswim" combines a measurement and a sonification module, and was developed in cooperation between BeSB Sound Engineering GmbH Berlin and the University of Hamburg. The system is designed as an easy-to-use measurement and real-time-audio feedback unit and is attached to the athlete's lower back. The IMU tracks 3D-acceleration (range ±2 g, accuracy 1%) and via on-board sensor fusion 3D-rotation (resolution 0.1°) with a sampling rate of 100 Hz, while the sonification module converts the data in real-time into sound and transmits it to the athlete via wired, waterproof in-ear headphones. An algorithm specially developed determines the relevant parameters for each swim stroke, including the UDK, the vertical hip roll and the vertical acceleration (up and down motion). Previously selected parameters, here the forward acceleration of the UDK, can be acoustically displayed to the swimmer in real-time. However, it is also possible to sonify other parameters. The transformation of the data into sound was realized by parameter-mapping, whereby data values are mapped to acoustic attributes or parameters of sound (Nees & Walker, 2009). Sofiswim's measurement values are smoothed with a 4th order 5 Hz Butterworth low-pass filter and mapped to frequencies between 110 and 1760 Hz (±2 octaves around the standard pitch 440 Hz). In order to achieve an aurally correct, i.e., logarithmic mapping to the frequencies, the processed values are first mapped to the corresponding MIDI (Musical Instrument Digital Interface) numbers 45 (note A2) to 93 (note A6). The conversion to frequencies is then done by:

$$f(t) = 440 \cdot 2^{\frac{m(t)-69}{12}}$$

with $m(t)$ being the MIDI number and $f(t)$ the resulting frequency at the time t. The resulting frequencies are used to modulate a square wave oscillator, so that an increase in forward acceleration leads to higher pitches of the sound. Due to the rhythmic downward and upward kick movement, the tone pitch increases and decreases to represent forward acceleration and to support the athlete to maintain the same movement rhythm. This version was called continuous sonification. With a delay of less than 10 ms between action and sound, the real-time feedback requirements are met (Schaffert, Engel, Schlüter & Mattes, 2019).

The difference between the continuous and discrete sonification is that continuous sonification provides the athlete with a continuous acoustic signal, while discrete feedback provides only one tone per cycle, which corresponds to the maximum acceleration during this cycle. The software detects the acceleration maximum and then calculates the corresponding frequency as described above. A sine wave oscillator with an according envelope (attack 100 ms, release 500 ms) is used to produce a short, precise sound. And to make the resulting sound sequence more harmonic, the calculated frequency is rounded to the next note of the corresponding major scale. Optionally, at each zero-crossing of the acceleration after the global minimum, a fixed tone (note A2, MIDI number 45, frequency 110 Hz) sounds which, together with the tone produced at maximum propulsion, provides a feedback on rhythm and frequency.

6.3.3.4 Testing

The results of first pilot tests with elite national swimmers revealed that it was possible to distinguish the different movement phases in the sound. It was also possible for them to relate the sound to their own movement while swimming with the sonification. This was possible for both sonifications (continuous and discrete). With both versions, the athletes can distinguish between movements with less acceleration (e.g., small hip amplitude) from those with larger acceleration (e.g., large hip amplitude) on the basis of the resulting tone or sound. The same was found for flexed (less acceleration) and stretched feet (large acceleration) and UDKs with a low (less acceleration) and high frequency (large acceleration).

The swimmers also stated that it was possible to recognize a decrease in kick frequency during one trial. With this information, they instinctively tried to maintain the same kick frequency as long as possible. Therefore, they favored the discrete sonification as the pause between two successive maximum tones conveyed the change in movement rhythm audibly better compared to the continuous sonification.

Theoretically the continuous sonification offers the advantage of receiving information even about the decelerating parts of the UDK. However, this was not confirmed from the swimmers yet since it was not possible for them to identify differences in the sound between the different ways of executing the UDK and the resulting decelerating effect. On the other hand, the discrete sound provides only one sound per cycle, informing about the maximum acceleration value of each cycle that is represented in tone pitch. With that, the athletes reported that it was easier to recognize small differences in maximum forward acceleration compared to the continuous sonification. A sound signal, generated during the decelerating part of the movement, neither affected the execution nor informed it feedback about effectiveness.

Moreover, the participants found the tones generated with the discrete sonification more comfortable compared to the sounds from the continuous sonification. Also, the discrete sonification provides information about changes in acceleration and movement rhythm, and was perceived as being more comfortable to listen to. However, the missing feedback about the decelerating parts was not disadvantageous for the athletes. Since the swimmers preferred the discrete sonification over the continuous sonification, this version seems more suitable for a long-time use in swimming training.

6.4 Conclusions

In combination with trends previously mentioned and considering the current trend in sports science to provide multimodal information presentation, acoustic feedback systems are a promising tool included in the training process as an alternative or in addition to existing systems. The powerful capacity of the auditory system to detect temporal patterns of periodicity and structure in acoustic information, an evolutionary necessity for meaningful acoustic information processing (as in speech for example), makes the auditory system an excellent system to neurally calculate and encode rhythmicity in sensory signal processing (McDermott & Oxenham, 2008, Pearce & Wiggins, 2012), allowing tightly-closed interaction loops in real-time applications. Synthetically produced acoustic information as acoustic feedback then is created using sonification whereby complex data structures are mapped to sound and sound in turn is used to monitor changes in data. This additional given acoustic feedback offers abundances of possible applications for monitoring and observing movements and detecting changes therein. Moreover, especially auditory feedback is considered effective for motor control and motor learning (Effenberg, 2005; Effenberg et al., 2007).

Nevertheless, it has been repetitively argued, that some of the existent sonifications are not pleasant to listen to because they are rather functional. In pursuit of a more aesthetical sound output, some examples trying to go beyond the pure mechanical parameter mapping of data to tone, e.g., in swimming (Seibert & Hug, 2013) and rowing (Dubus, 2012). Also, authors sought solutions using aesthetics of music, using theory and elements (Godbout & Boyd, 2012) that inform users to e.g., know how far they are in the exercise, dependent on how far they are in the completion of a chord (Newbold, Bianchi-Berthouze & Gold, 2017). Even though data-based aesthetic sounds were provided here, the actual effectiveness of these sounds could not be verified.

This balancing act between functional information presentation with pure tones and data-based musical/instrumental information feedback with a more melodic sound experience is still under discussion. Hermann, Hunt and Neuhoff (2011) make a clear distinction between sonification and musical instruments based on the nature of the interaction: they argue that musical instruments should not be considered interactive sonification devices because their primary function is to transform human gestures into sound in order to express themselves. In contrast, interactive sonification systems transform data into sound (modulated and controlled by human movements) for the purpose of data analysis.

However, despite this separation, there are also examples that represent a combination of both approaches, in which the boundaries of this distinction are pushed, in that the sonification of movements to sound not only support the moving person (e.g., dancer) in their own perception of movement, but at the same time also in the expressivity of movement through musical accompaniment (Brown, 2019).

Based on this, new trends and topics have emerged in the last decade with new projects that have expanded the boundaries of what sonification is in the field of sports and rehabilitation. This opens the possibility in the future to identify new directions that could benefit the users of sonification in sports and beyond.

6.5 Future Directions

In future approaches, the boundaries between sonification and music in the context of sports will continue to blur. Although the traditional purely functionally oriented approach, in which sonification systems in sports function mainly as training tools and support athletes in competition, remains the most widespread application case for sonification, the further development and expansion in this field in various directions cannot – and should not – be ignored. This expansion ranges from pedagogy in the context of dance schools (Françoise et al., 2014; Giomi & Fratagnoli, 2018), enhancement of the spectator experience (Clay et al., 2012; Savery, Ayyagari, May, & Walker, 2019), motivation of non-athletes to exercise (Yang & Hunt, 2015), enhancement of the athlete's experience to make it richer or more enjoyable (Fan & Topel, 2014; Tajadura-Jiménez, et al., 2019), sport movement to create music and aesthetic sound (Brown & Paine, 2015; Hug, Seibert & Cslovjecsek, 2015) to social interaction (Mueller et al., 2012).

Consistently taken further, this means that the boundaries between sport, play and fun, as differentiated by Höner in 2011, are also beginning to blur. This in turn opens up new sonification approaches not only to support the athlete, but also to enhance the experience for a broader and more diverse audience, as described above.

As suggested by van Rheden, Grah, and Meschtscherjakov (2020), the future goals of sonification research may not just be to improve athletic performance, but to improve the athlete's experience. In the authors' vision, future systems will sonify a variety of sensor data to enhance this experience for both the athlete and the audience.

References

Anderson, R. (2010). Augmented feedback–the triptych conundrum. In *ISBS-Conference Proceedings Archive, 28*, 95–96.

Baudouin, A., & Hawkins, D. (2002). A biomechanical review of factors affecting rowing performance. *British Journal of Sports Medicine, 36*(6), 396–402.

Bigand, E., Lalitte, P., & Tillmann, B. (2008). Learning music: Prospects about implicit knowledge in music, new technologies and music education. In *Sound to Sense, Sense to Sound. A State of the Art in Sound and Music Computing*, 47–81. Berlin: Logos Verlag.

Brock, H., & Ohgi, Y. (2016). Towards better measurability-IMU-based feature extractors for motion performance evaluation. In *Proceedings of the 10th International Symposium on Computer Science in Sports (ISCSS)*, 109–116. Cham: Springer.

Brown, C. (2019). Machine tango: An interactive tango dance performance. In: *Proceedings of theThirteenth International Conference on Tangible, Embedded, and Embodied Interaction*, 565–569.

Brown, C., & Paine, G. (2015). Interactive tango milonga: Designing internal experience. In *Proceedings of the 2nd International Workshop on Movement and Computing*, 17–20.

Clay, A., Couture, N., Nigay, L., De La Riviere, J. B., Martin, J. C., Courgeon, M., ... & Domengero, G. (2012). Interactions and systems for augmenting a live dance performance. In *IEEE International Symposium on Mixed and Augmented Reality-Arts, Media, and Humanities (ISMAR-AMH)*, 29–38. IEEE.

Colman, V., Persyn, U., & Ungerechts, B. (1998). A mass of water added to the swimmer's mass to estimate the velocity in dolphin-like swimming below the water surface. In *VIII International Symposium-Biomechanics and Medicine in Swimming-Programme and Abstracts*, 49. Gummerus Printing.

Cossor, J., & Mason, B. (2001). Swim turn performances at the Sydney 2000 Olympic Games. In R. H. Sanders (Ed.), *ISBS-Conference Proceedings Archive*, 65–69. San Francisco, USA.

Dubus, G. (2012). Evaluation of four models for the sonification of elite rowing. *Journal on Multimodal User Interfaces, 5*(3), 143–156.

Effenberg, A. O. (2005). Movement Sonification: Effects on perception and action. *IEEE Multimedia, 12*(2):53–59.

Effenberg, A. O., Weber, A., Mattes, K., Fehse, U., & Mechling, H. (2007). Motor learning and auditory information: Is movement sonification efficient? *Journal of Sport & Exercise Psychology, 29*, 66.

Fan, J., & Topel, S. (2014). Sonic Taiji: A Mobile Instrument for Taiji Performance. Georgia Institute of Technology.

Ferguson, S., Martens, W. L., & Cabrera, D. (2011). Statistical sonification for exploratory data analysis. In *The sonification handbook*.

Fontana, F., Morreale, F., Regia-Corte, T., Lécuyer, A., & Marchal, M. (2011). Auditory recognition of floor surfaces by temporal and spectral cues of walking. *International Community for Auditory Display*.

Formaggia, L., Mola, A., Parolini, N., & Pischiutta, M. (2010). A three-dimensional model for the dynamics and hydrodynamics of rowing boats. In *Proceedings of the Institution of Mechanical Engineers, Part P: Journal of Sports Engineering and Technology, 224*(1), 51–61.

Françoise, J., Alaoui, S. F., Schiphorst, T., & Bevilacqua, F. (2014). Vocalizing dance movement for interactive sonification of Laban Effort Factors. In *Proceedings of the Conference on Designing Interactive Systems: Processes, Practices, Methods, and Techniques, DIS*, 1079–1082. 10.1145/2598510.2598582

Giomi A., & Fratagnoli, F. (2018). Listening Touch: A Case Study about Multimodal Awareness in Movement Analysis with Interactive Sound Feedback. In *Proceedings of the 5th International Conference on Movement and Computing (Genoa, Italy) (MOCO '18), 4*, 1–8. New York, USA: Association for Computing Machinery. 10.1145/3212721.3212815

Godbout, A., & Boyd, J. E. (2012). Rhythmic sonic feedback for speed skating by real-time movement synchronization. *International Journal of Computer Science in Sport (International Association of Computer Science in Sport), 11*(3).

Hermann T. (2008). Taxonomy and definitions for sonification and auditory display. In *Proceedings of the 14th International Conference on Auditory Display (ICAD)*, 1–8.

Hermann, T. (2020). Sonification – A definition. Retrieved Nov 25, 2020 from https://sonification.de/son/definition/

Hermann, T., Hunt, A., & Neuhoff, J. G. (2011). *The sonification handbook*, 399–425. Berlin: Logos Verlag.

Höner, O., Hunt, A., Pauletto, S., Röber, N., Hermann, T., & Effenberg, A. O. (2011). Aiding movement with sonification in "exercise, play and sport". In *The sonification handbook*.

Hug, D., Seibert, G., & Cslovjecsek, M. (2015). Towards an Enactive Swimming Sonification: Exploring Multisensory Design and Musical Interpretation. In *Proceedings of Audiomostly 2015, 10th Conference on Interaction with Sound*.

James, D. A., Moritz, E. F., & Haake, S. (2006). The application of inertial sensors in elite sports monitoring. In E. F. Moritz & S. Haake (Eds.), *The engineering of sport 6*, 289–294. New York: Springer.

Kennel, C., Streese, L., Pizzera, A., Justen, C., Hohmann, T., & Raab, M. (2015). Auditory reafferences: The influence of real-time feedback on movement control. *Frontiers in Psychology, 6*(January), 69. 10.3389/fpsyg.2015.00069

Kleshnev, V. V. (2012). Power transfer between rowers through the boat. *Rowing Biomechanics News, 132,* 3–7.

Kleshnev, V. (2020). *Biomechanics of rowing: A unique insight into the technical and tactical aspects of elite rowing.* The Crowood Press.

Lahav, A., Saltzman, E., & Schlaug, G. (2007). Action representation of sound. *J Neurosci, 27*(2), 308–314.

Liebermann, D. G., Buchman, A. S., & Franks, I. M. (2006). Enhancement of motor rehabilitation through the use of information technologies. *Clin Biomech, 21*(1), 8–20.

Magill, R. A. (2011). Instruction and augmented feedback. In R. A. Magill (Ed.), *Motor learning and control,* 332–368, 9th Edition. New York: McGraw-Hill.

Maglischo, E. W. (1993). *Swimming even faster.* McGraw-Hill Humanities, Social Sciences & World Languages.

McDermott, J. H., & Oxenham, A. J. (2008). Music perception, pitch, and the auditory system. *Current Opinion in Neurobiology, 18*(4), 452–463.

McGookin, D., & Brewster, S. (2011). *The sonification handbook.*

Menzer, F., Brooks, A., Halje, P., Faller, C., Vetterli, M., & Blanke, O. (2010). Feeling in control of your footsteps: Conscious gait monitoring and the auditory consequences of footsteps. *Cognitive Neuroscience, 1*(3), 184–192. 10.1080/175 88921003743581

Mueller, F., Vetere, F., Gibbs, M., Edge, D., Agamanolis, S., Sheridan, J., & Heer, J. (2012). Balancing exertion experiences. In *Proceedings of theSIGCHI Conference on Human Factors in Computing Systems,* 1853–1862.

Nees, M., & Walker, B. N. (2009). Auditory Interfaces and Sonification. In Stephanidis C. (Eds.), *Universal access handbook,* 507–522. CRC Press.

Newbold, J. W., Bianchi-Berthouze, N., & Gold N. E. (2017). Musical expectancy in squat sonification for people who struggle with physical activity. In *Proceedings of the International Conference on Auditory Display (ICAD2017).* Georgia Institute of Technology.

Novais, M. L., Silva, A. J., Mantha, V. R., Ramos, R. J., Rouboa, A. I., Vilas-Boas, J. P., … & Marinho, D. A. (2012). The effect of depth on drag during the streamlined glide: A three-dimensional CFD analysis. *Journal of Human Kinetics, 33,* 55–62. 10.2478/v10078-012-0044-2

Pearce, M. T., & Wiggins, G. A. (2012). Auditory expectation: The information dynamics of music perception and cognition. *Topics in Cognitive Science, 4,* 625–652.

Rocchesso, D., & Serafin S. (2009). Sonic interaction design. *Int J Hum–Comput Stud, 67*(11), 905–906.

Savery, R., Ayyagari, M., May, K., & Walker, B. N. (2019). *Soccer sonification: Enhancing viewer experience.* Georgia Institute of Technology.

Schaffert, N., Engel, A., Schlüter, S., & Mattes, K. (2019). The sound of the underwater dolphin-kick: Developing real-time audio feedback in swimming. *Displays, 59,* 53–62.

Schaffert, N., & Gehret, R. (2013). Testing different versions of functional sonification as acoustic feedback for rowing. In *International Conference on Auditory Display (ICAD),* 331–335. Poland.

Schaffert, N., Janzen, T. B., Mattes, K., & Thaut, M. H. (2019). A review on the relationship between sound and movement in sports and rehabilitation. *Frontiers in Psychology*, *10*, 244.

Schaffert, N., & Mattes, K. (2014). Testing immediate and retention effects of acoustic feedback on the boat motion in high-performance rowing. *Journal of Human Sport & Exercise. 9*, 616–628. 10.14198/jhse.2014.92.02

Schaffert, N., & Mattes, K. (2015). Effects of acoustic feedback training in elite-standard para-rowing. *Journal of Sport Sciences. 33*, 411–418. 10.1080/02640414.2014.946438

Schaffert, N., & Mattes, K. (2016). Influence of acoustic feedback on boat speed and crew synchronization in elite junior rowing. *International Journal of Sport Science and Coaching. 11*, 832–845. 10.1177/1747954116676110

Schaffert, N., Mattes, K., Barrass, S., & Effenberg, A. O. (2009). Exploring function and aesthetics in sonifications for elite sports. In *Proceedings of the 2nd International Conference on Music Communication Science (ICoMCS2)*, 83, 86. HCSNet.

Schaffert, N., Mattes, K., & Effenberg, A. O. (2011). An investigation of online acoustic information for elite rowers in on-water training conditions. *Journal of Human Sport & Exercise*, *6*, 392–405. 10.4100/jhse.2011.62.20

Schmidt, R., & Lee, T. (2019). *Motor learning and performance. From Principles to application.* 6th Edition. Human Kinetics Publishers.

Schmidt R. A., & Wrisberg C. A. (2000). *Motor control and performance: A problem-based learning approach.* 2nd Edition. Champaign: Human Kinetics.

Seibert, G., & Hug, D. (2013). Bringing musicality to movement sonification: Design and evaluation of an auditory swimming coach. In *ACM International Conference Proceeding Series.* 10.1145/2544114.2544127

Tajadura-Jiménez, A., Newbold, J., Zhang, L., Rick, P., & Bianchi-Berthouze, N. (2019). As light as you aspire to be: Changing body perception with sound to support physical activity. In *Proceedings of the 2019 CHI Conference on Human Factors in Computing Systems,* 1–14.

van Rheden, V., Grah, T., & Meschtscherjakov, A. (2020). Sonification approaches in sports in the past decade: A literature review. In *Proceedings of the 15th International Conference on Audio Mostly*, 199–205.

Van Vugt, F. T., & Tillmann, B. (2014). Thresholds of auditory-motor coupling measured with a simple task in musicians and non-musicians: Was the sound simultaneous to the key press? *PloS One, 9*, e87176.

Wulf, G. (2007). Attentional focus and motor learning: A review of 10 years of research. In Hossner E.-J., Wenderoth N. (Eds.), *Gabriele Wulf on attentional focus and motor learning [Target article].* E-Journal Bewegung und Training *1*, 4–14.

Wulf G., & Prinz, W. (2001). Directing attention to movement effects enhances learning: A review. *Psych Bull Rev 8*, 648–660.

Wulf G., & Su, J. (2007). An External Focus of Attention Enhances Golf Shot Accuracy in Beginners and Experts, Research Quarterly for Exercise and Sport, *78*:4, 384–389, DOI: 10.1080/02701367.2007.10599436

Yang, J., & Hunt, A. (2015). *Real-time sonification of biceps curl exercise using muscular activity and kinematics.* Georgia Institute of Technology.

7 Augmented Reality for Sports Spectating and Coaching

Stefanie Zollmann, Tobias Langlotz,
Holger Regenbrecht, Chris Button,
Wei Hong Lo, and Steven Mills

7.1 Introduction

Spectator sports like football, rugby, and cricket have significantly changed the way people watch the games. Fifty years ago, the main option would have been to go to a sports field or stadium to watch the games live. Some more ambitious spectators would have brought their transistor radios to follow sports commentators while watching the game in parallel. In the last decades, many sports have become highly professionalized, stadium capacities were quickly reached, costs for watching games on site increased, media technologies developed, and more and more people are now watching sports events in front of a television set or on portable devices. On one hand, this allowed for different and closer viewing angles, and for the provision of additional, overlaid information display. On the other hand, "couch participation" lacks the atmosphere of a real stadium environment.

In 2020, many fans were excluded from attending live sports and forced into virtual observation because of the global coronavirus pandemic. Interestingly, the hiatus in stadium spectatorship has required sports broadcasters to rethink how they can synthesize the typical experience – e.g., by adding crowd noise – with varying levels of success (Majumdar & Naha, 2020). Many sports clubs and their loyal fan bases have lamented the absence of fans at live sport and the lack of emotion associated with 'crowd-less' broadcasted matches[1]. These recent experiences have clearly illustrated the widening gap that has appeared between the experiences of watching live sport at a venue and watching broadcasted coverage. It would appear that a new age of fandom is rapidly emerging and that sports and broadcasters alike need to seriously consider how best to attract and engage spectators in the future.

In addition, advancements in sports broadcasting and sports media in data capture have allowed broadcasters to retrieve and analyze sports-related information in real-time or near real-time. This information can include team statistics such as ball possession or scores and increasingly include player specific data such as running distances, fouls, and (more recently) heart rate and other physiological readings. Also, the access to slow-motion replays and digital animations is becoming more prevalent.

DOI: 10.4324/9781003205111-7

Spectators want commentators to review important refereeing decisions or crucial moments in a game from numerous angles and perspectives. This information is processed and visually presented in what has become a key component of sports broadcasting. These developments are ongoing and the next evolution of sports broadcasting is already emerging – Interactive sports broadcasting. This evolution is driven by advances in sensing technology and by the emerging paradigm of Visual Computing which combines Image Processing, Computer Vision, Visualization, Human-Computer-Interaction, and Computer Graphics. Thanks to these advances spectators can nowadays experience a soccer game from the perspective of a player (Rematas, Kemelmacher-Shlizerman, Curless, & Seitz, 2018), listen to live dialogue between a rugby referee and players or take a virtual seat in an America's Cup boat without leaving our homes[2]. However, spectators at live sporting events often miss out on this augmented live information. It is only available to remote viewers through broadcast media. This might affect the attractiveness of attending live sports events.

In this chapter, we will discuss the potential and the challenges of using Augmented Reality (AR) as an interface to access sports-related content on-site for sports spectating and coaching. We will discuss how AR as technology can improve the experience for live sports spectators as well as different kinds of sports spectators (e.g., fans, coaches, officials) could benefit from AR. AR has the potential to augment spectators' views with interactive real-time information and to bridge the widening gap between experiencing live sports events and remote interactive sports broadcasting. The development of mobile AR techniques that precisely overlay digital information onto the on-site spectator's or observer's view could help to increase the attractiveness of attending future live sporting events. AR has the potential to connect producers of sports statistics and event-related information with new audiences (i.e., spectators who currently choose to stay at home go to events and get sports statistics and event-related information and the live experience; coaches, managers, and trainers can view training sessions augmented with real-time and historic and statistical data while observing the performance of the athletes).

To develop the potential of AR as an interface to sport event-related data, it is important to understand and investigate novel visualization techniques for AR and vision-assisted tracking techniques. In this chapter, we will discuss Visual Computing techniques required to precisely overlay real-time sport-related information onto the on-field action. Vision-based tracking techniques in combination with adaptive visualization methods and local broadcasting of event data to the mobile devices of spectators have the potential to place rich information about sporting events in context for on-site spectators based on their individual location within the venue. In addition to solving computer vision and visualization challenges, there is also a need for a flexible infrastructure for on-site broadcasting of relevant sport-related information to spectators' mobile devices for an on-site mobile AR experience. Through this infrastructure, spectators at live events will be able to receive real-time, sport-related information as well as experience the atmosphere and drama of live sport as it unfolds.

Such an infrastructure will provide additional broadcasting outlets for digital content providers and thus will not only have an impact on spectator experience but also on the content creation industry. AR also has the potential to contribute to a more attractive proposition for sports venues and sports teams. Sports teams significantly benefit from increased numbers of spectators not just financially but also in terms of performance and building community and loyalty (Ludvigsen & Veerasawmy, 2010; Madrigal, 2006; Russell, 1983). This has been vividly demonstrated as sports were disrupted by the coronavirus pandemic and had to consider alternate means to engage their spectators. There are over 5000 major stadiums in the world[3], 300 of which are football (soccer) stadiums with seating capacities exceeding 40,000 each[4] creating a potential spectator base of 17 million users just for football (soccer) alone. When rugby, cricket, track-and-field or spill-over effects to other sports and non-sports-related events – the potential reach of sports stadia easily balloons out to 100 million or more spectators. All these spectators are paying customers and even if only a fraction of the ticket prices can be targeted this is an attractive market for content providers, venue managers, and potential mobile app developers and distributors. While in this chapter we mainly focus on professional (elite) sports events that take place inside sports stadia, there is also potential for such technology for recreational level team sports and beyond into other entertainment domains.

We will first discuss technical requirements and challenges to provide a solid foundation about what is required to create AR applications for on-site spectators based on their location within the venue. We will then discuss options for integrating these AR applications into a flexible infrastructure for on-site broadcasting that aggregates information from different sources and sends it to the spectators' mobile devices. Here we will apply novel techniques for precisely tracking mobile devices within the venue allowing us to place the aggregated information in the spectators' view. Spatial filters and contextual knowledge will help to preserve on-field action while complementing it with relevant information. The outcome of this research will be a new experience for the spectator and a more attractive proposition for the venue. Immediate beneficiaries of this research will be sports spectators and participants, digital content providers and both indoor and outdoor sports venue providers.

7.2 Augmented Reality

AR interfaces extend our view of the real-world environment by visually integrating digital information. With this, we can display additional information that is not physically present. For instance, this digital information can represent non-existing objects, meta-information, or hidden information.

In 1997 Azuma identified three major characteristics that describe an AR interface:

1. Combines real and virtual
2. Interactive in real-time
3. Registered in 3-D (Azuma, 1997)

Back at the time when Azuma formulated these requirements, it was still computationally expensive to implement these characteristics within one system. Expensive and powerful computing devices were required to combine them in one system. This limitation prevented the ubiquitous deployment of AR for conventional users due to high cost, bulky equipment, and limited availability. In recent years, with the increasing computational power of even small devices, omnipresent hardware, such as mobile phones and tablet computers, has grown powerful enough to fulfill Azuma's requirements. These further developments have worked towards a ubiquitous experience of the mixture of physical and virtual information and opened new fields of application, such as entertainment or advertisement, but also various professional applications.

While for decades AR was mainly driven by academic research, recently major investments from the industry have increased the availability of AR toolkits and frameworks. This allows developers with no prior AR expertise to develop AR applications. The reduction of entry requirements to AR development has contributed to many new AR applications that have previously only been described in research work. Examples include AR applications for furniture shopping (e.g., IKEA Place), games combining virtual and digital elements (e.g., Pokémon Go or Lego), or even AR-supported measuring tools. Most of the applications focus on small-scale environments, usually indoor locations. However, lot of application areas are there beyond these physical environments that would highly benefit from AR. For example, outdoor trampers and climbers are increasingly using AR topographical mapping technology to help them preview and plan their routes (Wiehr, Daiber, Kosmalla, & Krüger, 2017).

In 2016/2017, Pokémon Go became a worldwide phenomenon when it was taken up by millions of users, illustrating the potential of AR applications to be used alongside mobile devices. Mobile AR today has the potential to change users' experience and appreciation of sporting events in the same way as adding computer graphics to television broadcasting did in the '90s. Instead of augmenting the spectator's living room, AR has the potential to bring digital information directly to a spectator at a sports event, enriching the quality and attractiveness of those events. This will foster new forms of content provision and delivery, leading to further innovation and growth.

Pokémon Go's success was based on a novel on-site experience and on the economy of scale of free-of-charge offerings. The success of AR for sports spectating and training will also be based on the quality of the experience, but also on the quality of the content delivery for paying customers. Spectators at sports events are willing to spend a major amount on tickets and are expecting a quality experience for their money. Super Bowl spectators in the US are already paying US$2500 on average per ticket. The quality of the experience has to match these increasing ticket prices.

Content providers for added-value sports broadcasting experiences are seeking new and innovative forms of delivery. AR as an interface for spectators will give them the opportunity to offer new forms of end-user experiences. They will be able to deliver these directly to end-users, making them less dependent on established TV and internet broadcasters. Multinational corporate companies like Intel want to disrupt the market by offering immersive virtual reality (VR) sports experiences for the living room (e.g., Project Alloy) requiring billion-dollar investments in capturing the live events. However, "You need to experience sports events live and they just aren't the same when you're not in the stadium" says Intel's CEO Brian Krzanich[5] indicating that VR in the living room alone might not completely replace stadium attendance.

7.3 Technical Requirements

AR comes with a set of technical requirements for allowing it to be embedded into a user's view. These requirements are closely related to the characteristics of AR interfaces set by Azuma (Azuma, 1997).

In this section, we discuss those in particular (1) tracking and registration methods, (2) how to combine virtual and real information as well as (3) interaction techniques in the context of sports spectating and coaching.

7.3.1 Registration

In order to visualize any digital content in an AR interface, the first prerequisite is having suitable registration and tracking techniques available. Registration in the context of AR describes how virtual objects are aligned to the real world and how to assure that the digital data is correctly placed in relation to real-world objects. Often, the registration can consist of a localization approach coupled with tracking technology. The localization approach thereby computes an initial spatial relationship between the user's device and the real world. Tracking technology then supports the process of continually estimating and tracking this spatial relationship when the user moves away from a known position.

In AR, there are different options to achieve a correct registration, varying from simple maker-based registration techniques (Kato & Billinghurst, 1999) to tracking methods that use natural-features (Wagner, Reitmayr, Mulloni, Drummond, & Schmalstieg, 2008) sensor fusion (Schall et al., 2009) to localization-based approaches (Ventura & Hollerer,

2012). All these technologies use different approaches to achieve the same goal; aligning the virtual data in relationship to the physical world for a coherent AR visualization.

Marker-based and natural-feature-target based techniques are often used for indoor AR applications. For outdoor applications, they are usually not an option due to larger working environments and environmental influences. To achieve a reliable registration in unknown outdoor environments, often more sophisticated sensors are integrated into the AR setup or sensor-fusion approaches are used (Schall, Zollmann, & Reitmayr, 2013).

However, when it comes to sports spectating, one of the main challenges is that the area where we want to place content is quite large and dynamic. This creates a major challenge for a lot of the traditional registration techniques as well as the ones that target large open outdoor areas. For computing and placing digital overlays in the spectator's field of view, we need to compute the spatial relationship between the spectator's view and the event site and often also with regards to the digital content. Suitable AR interfaces are mobile phones or AR glasses (e.g., MS HoloLens[6]). To track the view we must solve two challenges: (1) The localization challenge in which we need to compute an accurate pose describing the position and rotation of the AR interface within the physical stadium. (2) Once this pose information is determined, we need to keep track of the movements of the AR interface with respect to the stadium. We call this the tracking challenge. Solving both, localization and tracking, allows us to register digital overlays in the spectator's view using the AR clients.

7.3.1.1 Localization

The main goal of the localization step is to compute the position and orientation of the AR interface (mobile phone or AR glasses) with regard to 3D models of the playfield or the venue. In our previous work, we investigated different methods for localizing a user within such an environment (Zollmann, Langlotz, Loos, Lo, & Baker, 2019). Options vary from user-guided methods to automatic localization methods.

7.3.1.1.1 USER-GUIDED LOCALIZATION

A user-guided localization method comes with the advantage that the user is in control of the registration technique and can provide immediate feedback. In addition, such methods have the advantage that they work independently from venues and do not require any additional dataset capture of the environment. However, they come with an additional workload for the user as the user needs to make sure they aligned the model accurately with the environment. During a lot of sports events, spectators are seated and have an allocated seat number, similarly, coaches or sports officials might select a predefined area with a good view over the playfield. If a seat number or a specific location is known beforehand, this can serve to provide a rough estimate of the user's position. Using this estimate we can provide an AR overlay of the sports venue which can be refined by the user by interactively aligning the overlay and sports ground. This approach requires a spatial mapping of all seats in the stadium which is also not always easy (e.g., not always numbered seats). If such a spatial mapping is not available another option is to use a traditional perspective-n-point solution that requires the user to align known 3D marks and 2D points in the AR client's view to compute the pose of the spectator's device. While both approaches produce suitable results, the usability is affected by the field of view of the AR client.

For instance, usually not all the corners of the sports field are visible to users, while other markings such as lines or advertisements are not always reliable (being badly visible or sometimes only roughly marked).

Besides that, sensor data of the mobile device can be used to specify a location on a map. The user is then required to align their view of the event site to the mapping interface. The success rate of this method depends heavily on the accuracy of the device's sensor data. Furthermore, GPS data is not always reliable in sports venues with large stands and roof structures. The disadvantage of the user-guided methods is that they put a major task load onto the user, and their ability to perform an accurate alignment will then, later on, have an impact on the user experience.

7.3.1.1.2 AUTOMATIC LOCALIZATION

In many cases, sensor-based localization (e.g., GPS) does not deliver the accuracy required as urban structures may affect the satellite-based locali-zation. Another option is through localization that uses a computer vision method. This approach uses known, fixed features from the environment and thus potentially has greater accuracy. An option for using such image features is given by the fact that a lot of sports venues display advertise-ments. These come with the advantage that they are very distinguishable from the rest of the environment and that they can be used for automatic recognition and pose estimation. So-called natural feature tracking targets use an approach where the device camera extracts image features and uses them for computing positioning information. Image features often represent 2D image points that are easily recognizable for computer vision algorithms such as corners or other prominent image regions. Popular choices for feature detection are the Scale-invariant Feature Transform (SIFT) or the more efficient Features from Accelerated Segment Test (FAST) (Rosten, Porter, & Drummond, 2010)). We then perform pose estimation by using those detected features and their counterparts in a reference image along with the known dimensions of the advertisements. AR Software Development Kits such as Vuforia[7] support natural feature tracking targets and make it easy to manage multiple targets at the same time. In our pre-vious work, we explored these options and obtained promising results for printed targets that are placed on the playfield. It is important to note that this approach strongly depends on the type and size of the advertisement and the position within the stadium and performance will be different if the used image target is far away or contains bad features according to SIFT or similar approaches.

For a lot of professional sports venues, there are 3D models of the environment available. In these cases, model-based approaches that use a 3D model of the environment for computing the spatial relationship between the device and playfield are an option. In our previous work, we investigated such methods with regards to their suitability for usage in AR (Baker, Zollmann, Mills, & Langlotz, 2019). The initial findings showed that these methods are location-dependent. Often, we find repetitive structures as well as dynamic elements that create difficulties to these model-based approaches. Alternatively, for several types of sports, it is possible to make use of the line-markings as a reference. This has been demonstrated for static broadcast cameras before and has also been used within AR interfaces (Skinner & Zollmann, 2019).

7.3.1.2 Tracking

The localization methods described in the previous section allow for an initial global alignment between the user's device and the real environment. After initialization, tracking methods allow for a continuous update of this relationship in order to compute the camera pose required for placing the AR content. Simultaneous Localization and Mapping (SLAM) approaches are often used as tracking technology in AR. SLAM is a tracking algorithm that has its origin in robotics and simultaneously computes a map of the environment while at the same time using it for localizing a device or camera (Davison & Murray, 2002). Within the mapping part of the algorithm often a 3D map is computed from 2D image features such as SIFT. These image features are identified and tracked in consecutive frames seen by a camera. Their 2D location in the image is then used to compute 3D rays and to compute the 3D position of those 2D image features using triangulation. The 3D position information is stored in a map that is then used for computing the positioning (and orientation) information of the device or camera. While traditional robotic SLAM approaches use stereo cameras or additional sensors (e.g., laser), for mobile AR applications monocular approaches have been developed to support single camera devices.

Modern SLAM approaches have shown promising results in larger-scale environments (Mur-Artal & Tardos, 2017) and proprietary AR SDKs such as ARKit[8] and ARCore[9] combine the data from motion sensors and SLAM approaches in Visual Inertial Odometry approaches for more robust AR tracking. We investigated some of these existing Visual Inertial Odometry approaches for their feasibility for AR in larger sports venues (Zollmann et al., 2019). Our initial tests turned out to work reliably enough for developing early prototypes and to demonstrate AR to sports spectators. However, we also noticed that the specific movement patterns for sports spectators (static position with mainly rotational movements) and location (large open environments with only minimal parallax) are conflicting with some assumptions traditionally made for SLAM trackers. In particular, the requirement of having a wide baseline between video frames for computing 3D information from 2D images via Structure from Motion (SfM) poses a challenge. A wide baseline allows triangulating image rays of known image features from two camera images. However, mostly stationary users within a large environment such as a stadium create the problem of not having enough translational movement for a large enough baseline. In order to address these challenges, we investigated the suitability of spherical SfM. For this purpose, we applied a spherical SfM method that computes the absolute pose based on a spherical movement constraint assuming all device motion can be represented by movements on a sphere. These approaches showed improvements when comparing them to traditional SLAM approaches. within the use case of a sports venue (Baker, Ventura, Zollmann, Mills, & Langlotz, 2020).

7.3.2 Visualization

Once we computed the exact camera position and viewing direction (camera pose) of the user or the device, we can use AR visualization techniques to overlay graphical representations of event-specific information into the field of view of the user. For a video see-through implementation using a mobile phone, a simple overlay of such information can be implemented by rendering a 3D model or annotation on top of the video image captured by the device's camera. For optical see-through displays such as the HoloLens, a 3D model would be displayed by rendering the 3D representation on the display. The combination of 3D content and view of the actual environment would then happen in the optical combiner of the device. For both, video see-through and optical see-through, the camera pose information is used to correctly align the 3D data with the view of the user's environment.

While such simple AR overlays are often used in various AR prototypes, they are often subject to perceptual issues (Zollmann et al., 2020). In particular, AR information visualization for complex environments such as sports venues is challenging since the environment, as well as the presented content is constantly changing. While content presentation in professional sports broadcasting relies on extensive manual editing to create a consistent visual output, this is not viable for on-site AR applications. There is often no control over the user's viewing perspectives, spectators and coaches will observe the player's action from various viewpoints and locations. Digital content placed on top of real-world objects might interfere with those real-world objects as well as with other digital elements. This is problematic as it is difficult to control the real environment where content is to be placed. In addition, the devices' screen sizes are often smaller compared to the devices that spectators use for consuming broadcast, e.g., larger TV screens.

These challenges can lead to cluttered presentations or occlusion of important parts of the view (e.g., placing an information label on top of a player). In order to address cluttered presentations in AR, research has been conducted previously on how image analysis and filtering can support automatic content placement and view management. For instance, image analysis techniques such as edge extraction and saliency computation have been used for placing content in AR application in a way to avoid overlapping items and occlusions as well as misalignment of digital items and real-world objects (Rosten, Reitmayr, & Drummond, 2005; Sandor, Cunningham, Dey, & Mattila, 2010; Zollmann, Grasset, Reitmayr, & Langlotz, 2014).

An option for addressing the challenges of small screen sizes is to make optimal use of the environment surrounding the user by using situated visualization techniques that analyze the user's environment for suitable content placement areas (Langlotz, Nguyen, Schmalstieg, & Grasset, 2014).

The visualization strongly depends on the type of input. In order to present sports event-related data to the user in an AR interface, different data sources have to be combined and transferred into a common 3D coordinate system. It is important that this coordinate system allows for putting the data in reference to the localization and tracking approaches. This is particularly a challenge as data sources can include 2D and 3D models (e.g., stadium models or static line overlays), information from player tracking that allows showing player paths or labels with information about a specific player. In addition, sports event-related information can also be provided by commercial sports databases sometimes referring to 2D or 3D coordinates, but sometimes also being related to a specific event or a team. In this case, it is important to define a spatial component relevant to the specific data. In our previous work, we demonstrated a system that integrates these different data sources for this purpose (Zollmann et al., 2019). In order to do so, we identified three main content categories for visualization in AR, these include **(a) Player-based, (b) Team-based, and (c) Game-based**. Examples for player-related data would be player names as well as physiological data such as heart rates. Team-based visualizations include visualization of data that corresponds to a specific team, such as ball possessions or field events. Game-based visualization includes the visualization of data that is relevant to the overall game or training event showing game rules, instructions, or hints such as offside that are not specific to a team or a player. In addition to these previous categories, an AR interface can also be used for visualizing crowd-based data. Crowd-based visualization could help to make the event more collaborative and share information between users on-site. This can include interactive games or entertainments for the crowd during breaks or even for emergency situations (e.g., evacuations) (Lo, Zollmann, Regenbrecht, & Loos, 2019).

In addition, data can be defined as spatially anchored. This means the data to be presented already comes with a spatial coordinate that allows the

AR interface to put it into the correct spatial location. Examples would be 3D models of game-related items, such as goalposts, or 2D representations of game-related objects such as line-markings. For training purposes, this could also include the rendering or a 3D crowd to create a more live event-like setting for the players during a training session.

One particular challenge relevant to using AR as an interface for sports spectating and training is that some information traditionally delivered for TV broadcasting comes in close to real-time (e.g., with a latency of seconds). This is no problem for TV production as there will be a small delay when sending the content, but the combination of information and footage is in sync. However, for AR, real-time overlays are required as the user is directly presented with their view of the environment. An offset between the action on the field and the presented data will be directly visible. To compensate for this and to adjust for delays in data delivery that are unavoidable, indirect AR can be a suitable solution. Indirect AR is an alternative implementation of an AR interface that uses a previously captured representation of the environment such as a panorama (Wither, Tsai, & Azuma, 2011). In contrast to direct AR where the overlay is shown on top of the live camera feed of the device, within the indirect AR interface, digital information can be overlaid on top of either a panoramic image, panoramic video, or a rendering of the scene. The advantage of this alternative interface is that one is able to replay a scene (thus relaxing the latency issue for some of the content) as we are not relying on the live camera feed while still keeping the immersiveness of the AR interface as the device is still fully tracked and responding to the users' movements(Zollmann et al., 2019). While this form of AR is only of limited value for replacing live augmentation of an entire event, it can be used for replays or the visualization of time-critical content.

7.3.3 Interaction

Once the localization, tracking, and visualization requirements are addressed, spectators can access event-related data on-site during a game or a training session. However, often there is a lot of different content to access, so one of the remaining questions is how will spectators interact with this content? While mobile AR interfaces such as a video see-through AR app on mobile allow for using standard user interface elements, head-worn implementations require alternative ways of interacting with the content. In particular, gesture-based and speech-based input has gained a lot of interest in recent years and has been integrated into commercial devices such as the HoloLens. In feasibility tests with our early prototypes, we noticed some usage patterns of mobile AR applications for sports spectating. In particular, that the AR interface would not be used continuously but rather in situations when users want to access specific information. Holding up a mobile phone for the whole duration of a match seems to be unfeasible. This is in contrast to the usage of head-worn devices. While these devices are still too bulky for casual users, they have the potential of better and more seamless integration of the interfaces as there is no explicit need for the user to switch on the device to look up any information. While speech input seems to be unfeasible for usage during large sports events, gesture input could still be an option for live sports events.

7.4 Opportunities of AR for Sports Spectating and Coaching

AR interfaces provide a set of benefits and potential for both sports spectators and to inform coaches for training. Sports spectators will benefit from the option of accessing data on-site in a similar way they are used to consuming content at home in front of the television. Sports spectators are already used to having overlays of graphics content on top of broadcast footage. Embedding a similar visual representation into their own perspective of the game on-site has the potential to deliver additional statistics and information that can be helpful for game understanding. This also has the potential to create more engagement and excitement during a sports event supporting fans but also sports teams. Instead of simply attending a match, AR provides a more interactive experience. Bringing digital content into the right spatial context has the potential to reduce the mental workload that is required when accessing additional data such as from traditional fan-focused mobile applications and web interfaces. Spectators can explore game-relevant statistics in the actual context of the action on the field. For instance, a user could tap on a player and a 3D label with player statics will appear next to the player, heat maps that visualize game relevant information will directly be overlaid on the pitch or team relevant statistics will appear on the site of each team. This will make it much easier to directly access relevant data.

Similarly, training staff and other professional personnel involved in sports team training and coaching will benefit from the ability to access data directly within the field of action. During a training session, a coach or manager could use AR as an interface to directly access additional stats about a player. Current session-related data such as events or historical data from previous matches could be displayed on top of the field action instead of using a 2D interface. If player tracking or ball tracking data is available this could be used to compare the performance of one player to another one. Players could use that to revise their own performance.

7.5 Conclusion

In this chapter, we discussed the requirements, challenges, and opportunities that arise from using AR as an interface for sports spectating and training. The main idea of AR as an interface to sports event-related data and content is to bring information to users on-site by overlaying it on top of their view of the actual environment. Examples are labels attached to players that display names and additional information such as physiological data or performances from previous events, or the visualization of game-related information such as heat maps or explanations relevant to the game development.

To use AR as an interface for sports spectating and training, several requirements need to be addressed such as tracking and localization, visualization, and interaction. In this chapter, we gave a brief introduction to AR and discussed solutions for addressing the main requirements. We also discussed the opportunities that arise when using AR as an interface for sports spectating and coaching. However, it is important to note that there are several remaining challenges. For instance, the success of AR interfaces is tightly related to the improvement of display technology – in particular, when it comes to head-worn displays, there is a need for reducing the weight and improving the acceptability of wearing hardware and sensors. There is a need for higher accuracy for localization and tracking to make sure that displayed content appears in the correct position. This is tightly connected to the development of better and additional sensors as well as the improvement of tracking algorithms. Also, current requirements with regards to power consumption are often a challenge for long-term use of AR interfaces, as the requirements for tracking algorithms and rendering often need high processing power. Using this technology for a complete sports event is still a challenge. In addition, social and ethical aspects need to be considered carefully, e.g., to address privacy concerns around and issues around continuous use of AR interfaces and with large crowds. Addressing all these challenges to provide an AR-enhanced experience for sports spectators remains a challenge but clearly not an insurmountable one.

Acknowledgements

We would like to thank Animation Research Ltd and in particular Sir Ian Taylor and John Rendall for their valuable input and support. We also thank the Forsyth Barr Stadium, the Highlanders and Otago Rugby (ORFU), for the opportunity to do on-site testing with our AR prototypes, and OptaPerform for their support with input data. In addition, we would like to thank Mike Denham and Craig Tidey from the School of Surveying at the University of Otago for their support in surveying the stadium model. The work discussed in this chapter is part of the ARSpectator project supported by an MBIE Endeavour Smart Ideas grant in New Zealand.

Notes

1 Premier League's home edge has gone in pandemic era: The impact of fan-less games in England and Europe (espn.com).
2 https://arl.co.nz/arl-news/221-on-board-the-america-s-cup
3 http://www.worldstadiums.com/
4 https://en.wikipedia.org/wiki/List_of_association_football_stadiums_by_capacity
5 https://www.si.com/edge/2017/01/11/future-virtual-reality-merged-sports-intel-ces-2017

6 HoloLens: https://www.microsoft.com/en-us/hololens
7 https://developer.vuforia.com
8 ARKit: https://developer.apple.com/augmented-reality/
9 ARCore: https://developers.google.com/ar

References

Azuma, R. T. (1997). A survey of augmented reality. *Presence: Teleoperators and Virtual Environments, 6*(4), 355–385. Retrieved from http://nzdis.otago.ac.nz/projects/projects/berlin/repository/revisions/22/raw/trunk/Master's

Baker, L., Ventura, J., Zollmann, S., Mills, S., & Langlotz, T. (2020). SPLAT: Spherical Localization and Tracking in Large Spaces. In 2020 IEEE Conference on Virtual Reality and 3D User Interfaces (VR), 809–817. 10.1109/VR46266.2020.00105

Baker, L., Zollmann, S., Mills, S., & Langlotz, T. (2019). Softposit for Augmented Reality in Complex Environments: Limitations and Challenges. In *International Conference Image and Vision Computing New Zealand* (Vol. 2018-November). 10.1109/IVCNZ.2018.8634761

Davison, A. J., & Murray, D. W. (2002). Simultaneous localization and map-building using active vision. *IEEE Transactions on Pattern Analysis and Machine Intelligence, 24*(7), 865–880. 10.1109/TPAMI.2002.1017615

Kato, H., & Billinghurst, M. (1999). Marker Tracking and HMD Calibration for a Video-Based Augmented Reality Conferencing System, 85. Retrieved from http://dl.acm.org/citation.cfm?id=857202.858134

Langlotz, T., Nguyen, T., Schmalstieg, D., & Grasset, R. (2014). Next-Generation Augmented Reality Browsers: Rich, Seamless, and Adaptive. *Proceedings of the IEEE. 102*(2), 155–169. 10.1109/JPROC.2013.2294255

Lo, W. H., Zollmann, S., Regenbrecht, H., & Loos, M. (2019). From Lab to Field: Demonstrating Mixed Reality Prototypes for Augmented Sports Experiences. In *Proceedings of the 17th International Conference on Virtual-Reality Continuum and its Applications in Industry (VRCAI '19)*. Article 62, 1–2, New York, NY, USA: Association for Computing Machinery. 10.1145/3359997.3365728

Ludvigsen, M., & Veerasawmy, R. (2010). Designing Technology for Active Spectator Experiences at Sporting Events. In *Proceedings of the 22nd Conference of the Computer-Human Interaction Special Interest Group of Australia on Computer-Human Interaction (OZCHI '10)*, 96–103. New York, NY, USA: Association for Computing Machinery. 10.1145/1952222.1952243

Madrigal, R. (2006). Measuring the multidimensional nature of sporting event performance consumption. *Journal of Leisure Research, 38*(3), 226–267. https://doi.org/Article

Majumdar, B., & Naha, S. (2020). Live sport during the COVID-19 crisis: Fans as creative broadcasters. *Sport in Society, 23*(7), 1091–1099. 10.1080/17430437.2020.1776972

Mur-Artal, R., & Tardos, J. D. (2017). ORB-SLAM2: An Open-Source SLAM System for Monocular, Stereo, and RGB-D Cameras. *IEEE Transactions on Robotics, 33*(5), 1255–1262. 10.1109/TRO.2017.2705103

Rematas, K., Kemelmacher-Shlizerman, I., Curless, B., & Seitz, S. (2018). Soccer on Your Tabletop. In *2018 IEEE/CVF Conference on Computer Vision and Pattern Recognition*, 4738–4747. 10.1109/CVPR.2018.00498

Rosten, E., Porter, R., & Drummond, T. (2010). Faster and better: A machine learning approach to corner detection. *IEEE Transactions on Pattern Analysis and Machine Intelligence, 32*(1), 105–119. 10.1109/TPAMI.2008.275

Rosten, E., Reitmayr, G., & Drummond, T. (2005). Real-time video annotations for augmented reality. *Advances in Visual Computing.* Retrieved from http://link. springer.com/chapter/10.1007/11595755_36

Russell, G. W. (1983). Crowd size and density in relation to athletic aggression and performance. *Social Behavior & Personality: An International Journal.* Retrieved from http://search.ebscohost.com/login.aspx?direct=true&db=s3h&AN=8651254& lang=pt-br&site=ehost-live

Sandor, C., Cunningham, A., Dey, A., & Mattila, V. V. (2010). An Augmented Reality X-Ray System Based on Visual Saliency. In *IEEE International Symposium on Mixed and Augmented Reality (ISMAR 2010), 27–36.* IEEE. Retrieved from http://ieeexplore.ieee.org/xpls/abs_all.jsp?arnumber=5643547

Schall, G., Wagner, D., Reitmayr, G., Taichmann, E., Wieser, M., Schmalstieg, D., & Hofmann-Wellenhof, B. (2009). Global Pose Estimation Using Multi-Sensor Fusion for Outdoor Augmented Reality. In *2009 8th IEEE International Symposium on Mixed and Augmented Reality, 153–162.* Orlando, FL, USA: IEEE. 10.1109/ ISMAR.2009.5336489

Schall, G., Zollmann, S., & Reitmayr, G. (2013). Smart vidente: Advances in mobile augmented reality for interactive visualization of underground infrastructure. *Personal Ubiquitous Comput., 17*(7), 1533–1549. 10.1007/s00779-012-0599-x

Skinner, P., & Zollmann, S. (2019). Localisation for Augmented Reality at Sport Events. In *International Conference on Image and Vision Computing New Zealand (IVCNZ),* 1–6. 10.1109/IVCNZ48456.2019.8961006

Ventura, J., & Hollerer, T. (2012). Wide-Area Scene Mapping for Mobile Visual Tracking. In *2012 IEEE International Symposium on Mixed and Augmented Reality (ISMAR),* 3–12. Atlanta, GA, USA: IEEE. 10.1109/ISMAR.2012.6402531

Wagner, D., Reitmayr, G., Mulloni, A., Drummond, T., & Schmalstieg, D. (2008). Pose Tracking from Natural Features on Mobile Phones. In *IEEE International Symposium on Mixed and Augmented Reality (ISMAR 2008),* 125–134. Cambridge, UK: IEEE. 10.1109/ISMAR.2008.4637338

Wiehr, F., Daiber, F., Kosmalla, F., & Krüger, A. (2017). ARTopos - Augmented Reality Terrain Map Visualization for Collaborative Route Planning. In *Proceedings of the 2017 ACM International Joint Conference on Pervasive and Ubiquitous Computing and Proceedings of the 2017 ACM International Symposium on Wearable Computers (UbiComp '17),* 1047–1050. New York, NY, USA: Association for Computing Machinery. 10.1145/3123024.3124446

Wither, J., Tsai, Y.-T., & Azuma, R. (2011). Indirect augmented reality. *Computers & Graphics, 35*(4), 810–822. 10.1016/j.cag.2011.04.010

Zollmann, S., Grasset, R., Reitmayr, G., & Langlotz, T. (2014). Image-Based X-Ray Visualization Techniques for Spatial Understanding in Outdoor Augmented Reality. In *Proceedings of the 26th Australian Computer-Human Interaction Conference on Designing Futures: The Future of Design,* 194–203. New York, USA: ACM. 10.1145/2686612.2686642

Zollmann, S., Langlotz, T., Grasset, R., Lo, W. H., Mori, S., & Regenbrecht, H. (2020), Visualization Techniques in Augmented Reality: A Taxonomy, Methods and Patterns. In *IEEE Transactions on Visualization and Computer Graphics, 27*(9), 3808–3825, 1 Sept. 2021. 10.1109/TVCG.2020.2986247.

Zollmann, S., Langlotz, T., Loos, M., Lo, W. H., & Baker, L. (2019). Arspectator: Exploring Augmented Reality for Sport Events. In *SIGGRAPH Asia 2019 Technical Briefs (SA '19)*, 75–78. New York, NY, USA: Association for Computing Machinery. 10.1145/3355088.3365162

8 Designing Augmented Ball-Based Team Games

Kadri Rebane and Takuya Nojima

8.1 Introduction

Being active and participating in sports is beneficial to human well-being. This is well known and also proven by research (Penado & Dahn, 2005). Being physically active reduces the risks of, for example, cardiovascular diseases, type 2 diabetes, hypertension, and obesity (Drinkwater, 1996; Pate et al., 1995). It also positively affects the muscles, joints, and bones (Vuori, 1998). However, the World Health Organization (*Physical Activity*, n.d.) statistics claim that insufficient physical activity is one of the leading risk factors for death worldwide. Globally one in four adults is not active enough, among adolescence this number is more than 80%. Additionally, only about 20% of jobs in the USA require a moderate amount of physical activity. In the early 1960s, about 50% of the jobs had that requirement. This means that the energy expenditure during work has lowered and it correlates with the overall weight gain seen in the population (Church et al., 2011). As many people have sedentary jobs, they can only be active during their time off. This leaves all the responsibility to be active to the individual and assumes that people will find meaningful, enjoyable, and appropriate ways to be active.

According to the flow theory by psychologist Mihalyi Csikszentmihalyi, people are most motivated and can concentrate when in flow state. Flow state is achieved when the task at hand has a good match between the participant's perceived ability and the challenge that the activity provides (Csikszentmihalyi, 2008). So all the physical activity where people participate should correspond to their current physical fitness level.

Spending time with friends, popularity, fitness/health, social status, sports events, and relaxation through sport creates motivation and wish to join sports (Kondrič et al., 2013). By design, team games create social connections and can provide a fun way to spend time. These factors are important when people decide about participating in the activity (Allender et al., 2006). Additionally, new habits are easier to take up when done with other people, as the commitment is stronger when it is shared (Duhigg, 2012). People are, by nature, trying to be their best in what they do. According to the Self-determination theory (Ryan & Deci, 1985), the factors that enhance their self-motivation and

DOI: 10.4324/9781003205111-8

personality integration are competence, relatedness (being social and connected to others), and autonomy (independence, personal achievements). This means that satisfying these needs, makes it easier for people to strive for their goals. Also, the following elements are considered essential to keep the enjoyment of team sports: playing as a team, being challenged, getting praised, playing time, and positive attitude (Visek et al., 2014). Enjoyment is also an essential factor to keep people playing sports.

However, in team game settings, it can be challenging to reach that kind of environment. The main reason is that it requires the presence of many people at the same place and at the same time with an objective that can be satisfied by doing team sports. Also, it is easy for team games to become unenjoyable if the team members have very different skill levels as the more skillful players would feel bored and the less competent players lack confidence in their skills. Research shows a 70% quitting rate in extracurricular sports activities among adolescents (Visek et al., 2014). The main reason for dropping out is that the activity became unenjoyable over time.

Balancing or adding handicaps in a game is an effective way to improve self-esteem when done hidden. Using the conventional method of assigning levels to players, on the other hand, has led to reducing relatedness in players and lower self-esteem (Gerling et al., 2014). There are various techniques related to hidden ways of balancing or adding handicaps among players in computer games. Computer games are played and enjoyed by many people, and game designers and researchers have identified methods for making the games appealing to players. For example, having defined roles makes the players identify themselves with the part, act like they were the represented player, and contribute to the overall playing enjoyment (Hefner et al., 2007). Another enjoyment mechanism in computer games is effectance: the perception of causal influence when players feel like their actions make a difference in the game world (Klimmt et al., 2007). Although computer games are good at engaging people and providing enjoyment, they are also sedentary activities, significantly different from traditional physical sports.

One way to make such traditional physical sports more appealing is to gamify those non-virtual activities. Gamification is defined as *the practice of making activities more like games to make them more interesting or enjoyable* (*GAMIFICATION | Meaning in the Cambridge English Dictionary*, n.d.). And it has been proven that it can be successfully applied to physical activities as well: *Our results suggest that gamification improves not only attitudes towards and enjoyment of exercise but also shapes behavior in terms of an increase in exercise activity* (Hefner et al., 2007).

As a subsection of gamification, an emerging field integrating the physical world with the digital one, is exergames. Exergames are defined as "*a combination of exertion and video games including strength training, balance, and flexibility activities. Exergaming is playing exergames or any other video games to promote physical activity*" (Oh & Yang, 2010). Another definition of exertion games is a fusion of technology, play, and body (*About – Exertion*

Games Lab, n.d.). These games often require certain systems specially developed for the game and can cover various purposes from training to physiotherapy.

Augmented sports can be seen as a section in exertion sport. They are usually based on physical activity and game elements are added to enrich the overall experience (Mueller et al., 2016).

In our study, we focus on already existing games as they are familiar to people and have survived the test of time. This means that the rules are thought through and acceptable to many people. As analog games, they are limited to what is possible in the physical world. Augmenting these games can result in a more exciting and fun playing experience and provides the freedom to add new rules and game elements to concentrate either on a more specific or diverse audience.

This chapter explores augmentation categories, presents a model for designing augmented games, and introduces our case study of augmented dodgeball, which focuses on players with varying experiences and backgrounds so they can enjoy being active together. In the case study, we use both virtual parameters and adjustments in physical space as a tool to incorporate sophisticated rules and characters for players to add tactics and teamwork to games, otherwise relying solely on physical skills. With our developed game devices, we also can avoid cognitive overload during the game so that players can focus on playing.

8.2 Augmentation of Sports

According to the Cambridge dictionary, the word "augmenting" means *the process of increasing the size, value, or quality of something by adding to it* (*AUGMENTATION | Meaning in the Cambridge English Dictionary*, n.d.). This section shows how to categorize augmentation, present a model for augmenting an activity and give an overview of different technological systems that can be used for realizing the augmentation.

8.2.1 Augmentation Categories

When looking at augmentation, we can see many different ways to add something to an activity or sport. The augmentations can take place either in the virtual space or in the physical world. We introduce four different categories, which can be used to augment sports. The four ways of augmenting the activity can be used independently or combined by taking advantage of all the different changes that each form of augmenting could bring to the experience. Table 8.1 summarizes the augmentation categories.

By understanding the augmentation categories, it helps to have an overview of what is done in this field, how incorporating technology and physical modifications can be and have been used to design a new user experience. In

Table 8.1 Augmentation categories

Augmented element	Description	Examples
Uniting virtual and real worlds	Using physical movement as an input for a goal set in the virtual world	Pokémon Go (*Pokémon GO*, n.d.), Augmented climbing wall (Kajastila et al., 2016)
Add or restrict a physical sensation/ ability in the game	Changing the core thing on how or where people can move or which senses to use	Wheelchair basketball (*Wheelchair Basketball*, n.d.), blind soccer (*What Is Blind Soccer | Warriors*, n.d.), Muscleblazer (Kishishita et al., 2019)
Add/replace a sense or sensation that usually is/is not present	Making the game about a sense that people do not have or that is not well developed, so equipment could aid the player with the missing sense or replacing one of the existing senses	Truffle hunting (*Truffle Hunting in Italy – Traditional Truffle Hunting and Eating*, n.d.), finding a way in a maze, in the dark using some guide device
Change the place and/or dimension where the activity takes place	By changing the environment, new movements and challenges are created. Or it also makes a too challenging movement easier	SUP (stand up paddling) yoga (*What Is SUP Yoga? | ISLE | Blog*, n.d.), Luna G ball (*Yanace on Twitter: "超人スポーツ Lunar G Ball*, n.d.)

our work, we concentrate on team games with a ball. We introduce how we think about augmenting team games in the following subchapter.

8.2.2 Augmenting Team Games

In our project, we investigated how to augment team games in the sense of adding value to the playing experience so that the player would feel welcome and engaged in the game. Our focus is on the player and their playing experience in team game settings. We found that when augmenting team games, we can divide the augmentation into three levels: game design, player experience, and environment design (Figure 8.1). Each level should answer to certain questions about game design and therefore help to create a new augmented game.

On the game design level, the questions to answer are:

• What are the shortcomings of the game/sport?
• What are the purposes of augmentation?

These questions should help to get a big picture of what kind of activity and outcome is desired from the augmented game.

Figure 8.1 Levels of augmentation.

On the player experience level, the question to answer is:

• What is the desired player response for each action in the game?

  It makes it easier to imagine the game's outcome by understanding what kind of payer behavior is desired for each action in the game. This helps to start discussions on the desired game actions and the probable effects they might have.

The environment design questions to answer are:

• Where is the game taking place?
• Which sense/ways of movement are allowed and encouraged in the game?

We found that discussing the augmentation on these three levels helps organize the augmented game design and focus on the big picture created on the first game design level.

  The next thing to do is that all the stakeholders in the game should be considered. The stakeholders are players, judges, helping/assisting staff, and the audience (spectators and observers).

  Players are usually the center of attention for the game. The game should emerge from them; they are the ones directly involved with it, and most of their energy and attention are used toward achieving the end goal of their game. When designing for the players, it is important to notice that their attention is limited. Also, the information presented to them should be limited by the things they need to know according to the game design.

  Judges should have a complete overview of the game and should be able to promptly do their duty in the game. It is necessary that the judges have a comprehensive overview of the game at any point and that they understand the game and its rules to the fullest.

  Helping/assisting staff are like helpers for the judges. They should have a clear overview of what is required of them, but the interface for them can be limited and may not include the whole picture if it is required by the game design.

The audience is also an important stakeholder to consider as the audience can produce new players for the game, increase the gaming satisfaction for the players, and promote the game. The audience can be divided into spectators (people who share the tension of the game, incorporate some kind of cheering) and observers (they only view the game, maybe from a distance or via video, audio, or (social) media link, and have no influence at all on the players.) It is important to notice that the audience interfaces should make the state of the game easily understandable and give an overview of the whole area of gameplay. The interfaces for the audience can be more complex as the audience in general would focus on watching the game.

Not all of the games include all of the stakeholders, but when designing the tools and technology of the augmented game version, all stakeholders who are present have to be taken into consideration.

8.2.3 Interaction with the Ball

As we concentrate on team games with a ball, the ball's movement in the game is the center of attention and the source of action and excitement. Thus, we can say that the human-ball interaction plays a central role in the game. When we talk about augmented games and the augmentation uses virtualization technologies, we can also have the ball move in the physical world, virtual world, or have some kind of a combination of both. As the ball's status and movement are essential, the stakeholders need to know them during the game. In most ballgames, it is necessary to know which player interacts with the ball at a particular moment and where the ball is in the field. In most cases, this information is crucial for being presented to the players and the audience somehow.

When designing the player-ball interaction and deciding between physical, virtual, or a mixed ball representation, the following considerations should be made:

Player safety: when the game has a physical ball, all possible trajectories of the ball should be taken into consideration when designing the technical systems of the augmented game. For example, players wearing expensive and still fragile HMD systems might pose a risk to the players' safety and the equipment when the device gets hit by the ball, and/or players collide with each other. Although the rules of the game might forbid some actions such as throwing the ball in specific ways and directions, designers must accept the existence of few players who try to outsmart the rules and find ways to fulfill the game goal. Such players' behavior and response in the game might differ from what is expected by the designer.

Virtual world representation: when the player-ball interaction is brought to the virtual world, all stakeholders should still have a clear understanding of how the game is progressing. Virtual world representation can be different for each stakeholder group or even by each stakeholder if fulfilling the augmented game purpose. When representing the virtual world, it should also consider the cognitive load. Showing such information may require a certain level of cognitive load. Thus it has to be considered how much information is needed by any given stakeholder during the game. Creating a heavy cognitive load with the virtual world representation may lead the game more strategic, but it will also lose in tempo. Having a fast-paced game with no virtual world representation, on the other hand, results in the heavy reliance on some specific physical skill, no balancing, and gradual progress, etc.

Immersing the virtual and physical world: to have a game in two different dimensions, actions in the physical world should produce reactions in the virtual world and vice versa. These action/reaction pairs are expected to be logical, as designed, as explained by the rules, and understood almost immediately. For example, you expect the ball to move toward the target direction when you throw it. Having a reliable action/reaction relationship helps players gain trust in the system and focus on the game. If the action/reaction system fails, the players start to doubt the system, which creates tension and arguments between players (and possibly other stakeholders too) and results in low playing satisfaction.

8.2.4 Technical Solutions for Augmentation

In this section, some technical ways of creating digital augmentation are discussed. Digital augmentation is the newest and the least investigated way of augmenting sports. In digital augmentation, the key to success is to create appropriate action-reaction interaction. Although a fast and real-time interface could be the goal of each such action-reaction pair, game design concepts and different ideas can be tested out also without reaching the fully automatic real-time interface level. The key concepts for doing that are to manage player expectations of the system and incorporate human judges and the Wizard of Oz testing method. The Wizard of Oz is a testing method of systems and/or interfaces that do not exist yet. Instead of the testing actual one, all the reactions expected by the system are provided by a human actor. This is a great way to get to know and test out user behaviors in the augmented game. The augmentation system can be divided accordingly:

- Systems for tracking
- Systems for notifications

The tracking systems are the ones that track relevant user input (action) and notification systems are the ones that display the corresponding change in the (virtual) game (reaction). Table 8.2 gives an overview of the augmentation systems and their purposes:

Table 8.2 Tracking and notification system purposes

	Tracking system	*Notification system*
Purposes	Tracking the state of the ball (held, bounced, etc.)	Player awareness of the game
	Which player interacts with the ball	Player awareness of his status in the game
	Ball position on the field	Audience awareness of the game
	Audience input to the game	Audience participation interface (if applicable)
	Judge/staff input tracking	Interface for judges (if applicable)
		Interface for helping staff (if applicable)

8.2.4.1 Tracking System

In team ball-based games, the system for tracking involves all the technical solutions considering the ball and player interactions and the state of the game. It handles all of the input from all the stakeholders mentioned above and processes the received data. The main tracking systems relevant to ball team games can be the following:

1. Tracking the state of the ball (is it held by the player, is it bounced, did it hit someone, how?)
2. Tracking which player interacts with the ball
3. Tracing if the ball is on the field (if applicable)
4. Tracing the audience input to the game (if applicable)
5. Tracking the judge/helping staff input and decisions during the game

In ball games tracking the ball is essential. It plays a key role in the outcome of the game. There are many technical solutions available for doing it. The choice depends on the requirements of the actual game or event. Essentially the main goal of the tracking system would be to tell the other parts of the game system about the state of the ball. There are two main ways to track the state of the physical ball: computer vision and using a set of sensors applicable to the exact need of the ball tracking in the game. Although a tracing camera can be thought of as a sensor as well, in recent years computer vision has evolved a lot and become a very powerful high technology field, other sensors in use rely on more basic physical input (for example sound, pressure, changes in the magnetic field). The tracking systems can be divided based on which unit of the system sends out the collected data that is used to identify the state of the ball. The sensors can be placed either on the playing field, on the players, or on the ball.

Table 8.3 Tracking system overview

Tracking system	criteria	comment
Computer vision	pros	Multiple objects can be tracked with one setup
		Easy to use in a decided constant place
		After the initial setup easy to work with
		Not much equipment
	cons	The high cost of equipment
		Not portable
		Long development time
		Long learning curb for beginners
Sensors on the field	pros	Cost-effective
		Good when user input might be required
		Accurate in small space
	cons	Needs field setup
		Might not be portable (depends)
		Might require user input
		Accuracy can be low in big areas
Sensors on players	pros	Portable
		Cheap developing
		Easily scalable
	cons	Players must be equipped
		Needs to be safe and robust
Sensors inside the ball	pros	Portable
		Easy to experiment with
		No on spot setup
		Easy to modularize
	cons	Robustness required already on prototype
		Robustness on the prototype can mean
		loss of accuracy or some visuals

Table 8.3 gives an overview of the tracking systems with their pros and cons. The tracking systems can be used separately or combined, depending on the exact requirements of the game and environment.

8.2.4.2 Notification Systems

Notification systems have a much longer background and are pretty common in our everyday lives. Essentially, a notification system gives information about the state of something or shows the change in a situation. Good example of notification systems are doorbells and the ringing sound of a phone. These systems let us know that somebody wishes to connect with us at that moment (a change in a situation). In augmented games, the notification is handling everything connected for displaying the game's current state to all stakeholders. It can be divided accordingly into:

1. Players awareness of the status of the game and fellow players
2. Players awareness of their status

3. Audience awareness of the status of the game
4. Audience participation interface (if applicable)
5. Interface for judges (if applicable)
6. Interface for helping staff (if applicable)

Each notification system should consider that when it displays the reaction of the action, it should be timely unless hiding information until later is part of the game's design. If the notifications of activities come at random times and/or do not correspond to the actions of what players are doing in the game, playing satisfaction will be easily lost, and players will also lose trust in the system. This, in turn, can result in an unsatisfying playing experience.

Another aspect of notification systems is that they should consider the amount of cognitive load the system requires and the player's acceptable level about the load.

The third consideration point when designing notification systems is the amount of information needed by each stakeholder during the game. Hiding and delaying as well as giving different amounts of information to different stakeholders can be part of the game design and enhance the overall playing experience.

8.3 Augmented Dodgeball

Augmented dodgeball game serves as a case study of the augmented team and ball-based games (Figure 8.2). The theoretical considerations and augmentation methods described in the previous chapter are analyzed and we show how they can be put into effect to design a new team ball game. Augmented dodgeball game has been evaluated with playtests (Rebane, Inoue, et al., 2021; Rebane et al., 2017). The hardware for playing has also been revised several times (Rebane, Hoernmark, et al., 2021) to provide a better playing experience for the players.

Figure 8.2 Players during an augmented dodgeball game.

8.3.1 Basic Design

The augmented dodgeball game is based on the dodgeball game played around the world mainly in elementary schools. The rules in different countries vary, our design is based on the standard rules used in Japan for playing dodgeball.

Dodgeball is a game played by two rival teams. Each team has infield and outfield players. Infield players can only move inside the field and outfield player(s) are positioned outside of the field but on the opposite side of their team. The game starts with both teams having one outfield player. During the game, the players throw a ball at each other. When a person gets hit by the ball, they become outfield players. Only the person who started as an outfield player can then go and play as an infield player. The playfield can be seen in Figure 8.3.

We started this project by observing the dodgeball game and identifying which parts of the game would be good to augment for creating a balancing mechanism between players of different skill levels (Nojima et al., 2015). We noticed that some players act more confident during the game from the initial observations of dodgeball play. They play more aggressively, attack the opposing team more, and take more risks when catching the ball. On the other hand, some players concentrate on ducking from the ball rather than trying to catch it and rarely attack the opposing team by throwing the ball at them. The rest of the players fall somewhere in between. Also, it was common to see the weaker players having less game time, as when they got hit, they were out of the game but could support remotely as outfield players. As the game progresses, the aggressively playing people also get hit and become outfield players.

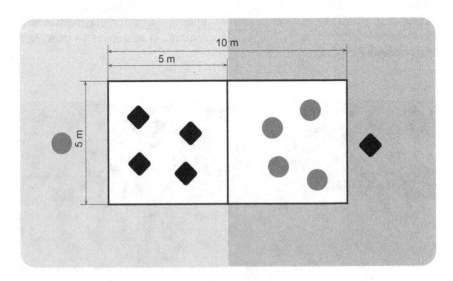

Figure 8.3 Dodgeball playing field layout. The dots represent players in team A and diamonds represent players in team B.

They also start dominating the game as outfield players, leaving the less aggressive players with little to no opportunity to contribute to the game. Research has shown that being able to contribute to the game is an essential part of the players' game satisfaction and a good indicator if the person would like to play again in the future (Visek et al., 2014).

Taking this phenomenon as a guide, we decided to develop augmented dodgeball as a more strategic game where all participants could concentrate on doing what they felt more comfortable at – attacking or ducking from the ball.

The main design goal of developing augmented dodgeball was to make a game that would enable people with different physical skill levels to play together and fruitfully spend time while being physically active. The design goals of the game are the following:

- Game level: increase collaboration; take attention to physical movement to tactics.
- Player experience: make all players feel that they are important; increase playing time (for weaker players).
- Environment design: appropriate field size.

Based on these answers, we decided to incorporate the following elements into the game:

- Add life-points to players so that they can survive for a longer time in the game.
- Create player roles with different players having different amounts of attack and defense points to promote balancing and collaboration in the game.
- Give players the freedom to choose their player roles, so that it is a voluntary action and players are not labeled based on their skills.
- For not increasing the players' cognitive load, the point calculation is done automatically, resulting in a virtual score in the game.
- We use a field sized 5 m × 10 m without any neutral zone. This field size was determined by playtests so that teams of 3–5 people could play comfortably. This means that the field would have enough space for all players to move. Also, it is large enough to be comfortable for throwing the ball at the other team.

As the augmented dodgeball has parameters that are not seen by the players and are quite difficult to calculate on the spot (parameters and their design is explained in the game parameter section), we can say that augmented dodgeball is a game taking place in two different dimensions: the physical one and the virtual one. The game takes place in the real world, with real players and a real ball. The players still try to hit other players with the ball as with traditional dodgeball, but when they get hit, they are not out of the game – instead, they lose life-points in the virtual world. The amount of life-

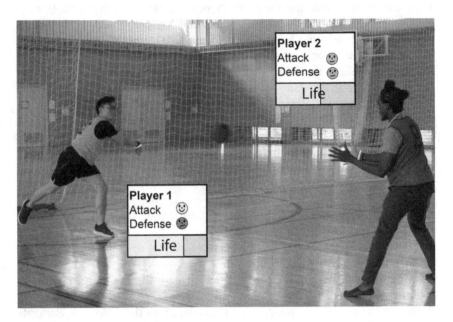

Figure 8.4 Augmented dodgeball concept.

points lost depends on both the attack power of the thrower of the ball and the defense power of the player who got hit. The concept of augmented dodgeball can be seen in Figure 8.4.

Using virtual parameters gives the option to add sophisticated game rules, which in the real world would make the game hard to play and follow as keeping track of the rules and points can easily get overwhelming. To incorporate the virtual game world with the physical one, visualization and game tracking systems are needed. In augmented dodgeball, we need to track who throws the ball and who gets hit and how many life-points each player has. For that, we have developed a game engine that manages all the information, thrower detection devices, and display devices that give the players information about the virtual parameters. Hit occurrence is entered into the system manually by a referee.

8.3.2 Game Parameters

In augmented dodgeball, players can play until they have life-points left. Additionally, they have attack and defense power. These virtual parameters were used to create player types that the players can choose from. For the moment, we have developed three player types: attacker, defender, and balanced. Attackers have high attack power but low defense power. Defenders have low attack power but high defense power and the balanced

Attacker **Balanced** **Defender**

Attack points 😊   Attack points 😐   Attack points 🙁

Defense points 🙁   Defense points 😐   Defense points 😊

Figure 8.5 Player types in augmented dodgeball.

type has medium attack and defense power. The player types with their logos can be seen in Figure 8.5.

To design the parameters for each character and to assure that they are appropriate: not too strong or too weak, we developed a game simulator. The simulator is based on observations made during an actual dodgeball match. Each game-action is assigned a time value and different skill levels are represented as a probability of hitting or getting hit by the ball. The simulator assigns each player with a random skill level and a random role. Then 10,000 games are simulated. The simulator outputs the average time required for the match, the number of people left on the field, and the team's number of wins. We were looking for parameters that would result in about 10 minutes per one game to keep the game exciting and avoid getting too tired. We also wanted to have as few people as possible left on the field at the end of the game to ensure the game would be interesting until the end. Additionally, we were looking for parameters that would result in both teams win evenly to ensure a good balance of parameters.

Based on these requirements, we found that the parameters in Table 8.4 result in good game metrics.

The points lost when getting hit are calculated by the following formula. The same formula is used in many video games.

Table 8.4 Player role parameters

Player role	Life-points	Attack power	Defense power
Attacker	120	140	120
Balanced	120	120	160
Defender	120	110	180

$$D = AP - \frac{DP}{2}$$

Where:

D – Damage to be reduced from life-points
AP – Attack power of the thrower
DP – Defense power of the person who got hit

The new life-points score is obtained when the D (damage) is subtracted from the old life-points score of the player who got hit.

8.3.3 Augmented Dodgeball System Requirements

Augmented dodgeball consists of physical and virtual layers. To realize the augmented dodgeball game, we need to know who throws the ball, who gets hit, have a point management system to keep track of the course of the game, and a way to notify players as well as the audience about the state of the game also on virtual layer. The augmented dodgeball system overview can be seen in Figure 8.6.

The center of the system is the game database which is used to hold all game parameters and the state of the game. It shares information with the game engine where the rules and logic of the game are stored. The database is taking input from the catch and hit detection systems and provides output data for the player and audience notification devices.

8.3.3.1 Notification Devices

Notification devices are used to present the game information to the players and the audience. In augmented dodgeball, two kinds of display devices are used. One is a scoreboard placed on the side of the playing field. It lists the players' life-points, player roles, which player is holding the ball and who is

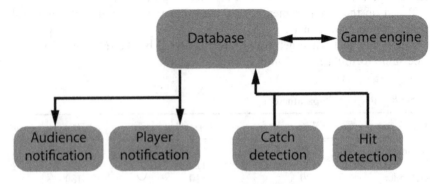

Figure 8.6 System diagram for augmented dodgeball.

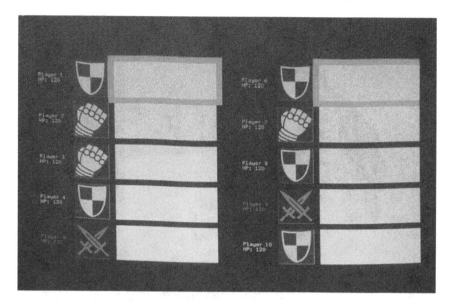

Figure 8.7 Augmented dodgeball scoreboard.

playing as an infield player, and who is an outfield player. The scoreboard can be seen in Figure 8.7. The light bars represent life-points each player has. Next to the bar is the player role logo and the brown frame indicates an outfield player. During the game, when a player is holding the ball, it is represented with a yellow frame and a ball on that player's life-points status bar. To distinguish player roles, all players wear a shirt corresponding to their role. The attacker players wear a red shirt, defenders a blue shirt, and balanced players wear a green shirt. In Figure 8.2, players can be seen while playing augmented dodgeball.

The second notification device is worn by the players and gives information about their current status on their wrists. This enables the players to see their status without turning their head onto the screen, which is placed on the side of the playing field. The wrist devices are also equipped with LED strips to enable other players to keep an eye on the remaining life-points of their peers. This information can be useful when deciding on some in-game actions. The device worn by the players can be seen in Figure 8.8. The device used is the M5 stack (*ESP32 GREY Development Kit with 9 Axis Sensor – M5stack-Store*, n.d.) which is a watch-type developing board. It is equipped with an ESP32 microcontroller, display, and has Wi-Fi capabilities.

8.3.3.2 Detecting Game Actions

In augmented dodgeball, getting hit by the ball is a core game element. When a player gets hit, they lose life-points based on their virtual defense

Figure 8.8 Device worn by players during the augmented dodgeball game.

power and the attack power of the player who threw the ball. To automate this calculation, the game engine needs information about who threw the ball and who got hit. The thrower information is passed to the system automatically by using a Hall sensor on the glove that the player is wearing and placing small magnets on the ball. When the player holds the ball, the sensor can detect the magnetic field and the wearable system sends out a signal to the database with the ID number of the player holding the ball. Hit occurrence is inserted into the database manually by a game referee.

Figure 8.9 demonstrates the glove with a Hall sensor worn by the players during the augmented dodgeball game.

8.3.4 Additional Game Elements

To further add balancing and excitement to the game and to add actions that players felt were needed, we have also experimented with several additional game elements.

8.3.4.1 Plus Mode

The normal mode is a mode where all the same types of players share the same parameters (for example, an Attacker in team A has the same defense and attack points as Attacker(s) in team B). The plus mode will be activated automatically when one team has established its superiority in the game. In

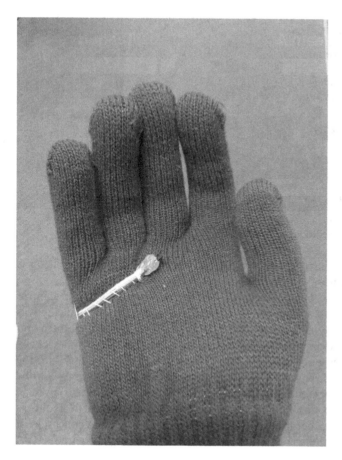

Figure 8.9 Glove with a Hall sensor used in the augmented dodgeball game.

the augmented dodgeball system, the mode is activated when one team has at least two players more in the field than the other team. In the plus mode, the attack and defense points of all the inferior team's players will increase by 10–20 points depending on their player role. This gives a slight advantage to the inferior team's players to encourage their playing motivation. Activation of the plus mode is also reflected on the scoreboard. In Figure 8.10, it can be seen that the logos of players 6 and 7 on the B team have changed color from green to yellow. This marks that the game is now in plus mode, and these players have increased attack and defense power.

8.3.4.2 Balancing in Two Layers

In augmented dodgeball, players can choose which virtual character they want to play in the game (attacker, defender, balanced). Players also know

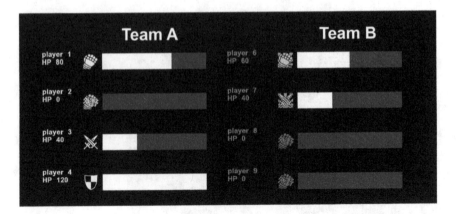

Figure 8.10 Augmented dodgeball scoreboard with plus mode enabled.

that the characters have different virtual parameters and which character is better in what skill.

To adjust also the balance between teams (for example when a team of adults is playing with a team of children), the second level of balancing was created. This is done hidden from the players in order to not label them and therefore the adjustments are not shown on the screen on the wrist device. At the beginning of the game, the referee has the option to individually adjust the virtual players of all players. This also allows adjusting the game time to be shorter or longer depending on how many life-points each player is given.

At the same time, the plus mode of the game was disabled to avoid labeling the teams as either leading or falling behind and to avoid confusion in players that found the concept hard to understand during the game.

8.3.4.3 Footswitch

The footswitch is a device comprised of pressure sensors and a small microcontroller and a wireless module. It was placed on the playing field and designed flat, so it would not be a tripping hazard for the player. At random times during the game, an audio signal would sound. The team whose player stands on the footswitch first would have increased attack power and defense power for the next 30 seconds of the game. This device would allow players to contribute to the game even when they are not handling the ball.

8.3.4.4 Spectator Participation System

We have also experimented with a spectator participation system. In this system, spectators could access a webpage and cast a vote for their favorite team. Every 5 minutes, the votes were counted and the team who got more votes had an increase in either of the attack power or defense power for the next 30 seconds. With this system, we hoped to increase the appeal of the game in the eyes of the spectators to promote the game. Moreover, for players, it provides an unexpected game element.

8.4 Discussion

We have demos of developed augmented dodgeball in many local and international events. In those demos, we could have chances to observe players' and spectators' activity and interview them. The following is what we learned by observing the players and from the player feedback after the games.

The possibility of a high-speed game makes the game attractive and exciting. However, when supporting hardware systems are not fully automatic or there are glitches in the operation, players can get frustrated very easily. That results in arguments and the overall game enjoyment suffers. Managing player expectations beforehand and explaining the game flow helps to do that. In our experiments and demos, players were very receptive to new rules and game flows. For example, explaining game pauses that are used to insert hit data into the system can be used as a time to organize player positions on the field and exchange some ideas about strategy. All player actions should be relevant in the game. When players are asked to do some action, it should be an action that makes sense in the game, not because the system requires it. While working on our first prototype with a helmet (Rebane et al., 2017), we instructed the players to touch the ball on the helmet to "equip it with power". Although this movement was sometimes forgotten, players understood it and were cooperative to do the additional movement.

As the augmented dodgeball takes place also in the virtual layer, it is essential to express that to players in a timely and easy-to-understand manner. The characteristics that do not change during the game are displayed through a static analog display (the color of players' shirts representing their player role). And for dynamically changing parameters (like life-points), technological solutions are used. In augmented dodgeball, we used both a personal wrist-watch type device and a scoreboard. About half of the players preferred a personal wrist-watch type device and the other half liked to see the status on the scoreboard. Among those players, we could find many children. They also enjoyed the game, but the excitement came from the colorful LEDs on the wristwatch-type device. They paid less attention to the augmented game rules. There are easier ways to create excitement about sports for children. We see that augmented sports has more potential to attract adults and adolescence into being physically active.

8.5 Conclusion

People are made to move. The benefits of physical activity are well known and backed by research. The advances in technology have altered the way lots of people work and livelihood can be made without much physical movement. As everyday life does not require a good level of physical fitness anymore, being active has increasingly become a hobby or a past-time activity. And not enough people take part in it. A good way to promote physical activity is to associate it with some human core needs. One of these is belonging to a group and interact with others. Research also confirms that social aspects of the activity play an important part in the decision of joining or not. This chapter introduced a way to think about enhancing team games and gave an overview on how to design the game and equipment to enhance the playing experience. Our case study with augmented dodgeball showed that using computer game elements to augment games can provide an interesting gaming experience, unite people with different skill levels and there is potential to enhance the communication between people with game design. Although it requires special equipment and sophisticated game design, this method could contribute to increasing participation in physical activity, especially among adults and adolescence. Augmented games can empower kids to be able to enjoy a game with adults but the kids do not need high technology to get excited about movement and sports.

References

About – Exertion Games Lab. (n.d.). Retrieved October 5, 2020, from http://exertiongameslab.org/about

Allender, S., Cowburn, G., & Foster, C. (2006). Understanding Participation in Sport and Physical Activity among Children and Adults: A Review of Qualitative Studies. *Health Education Research*, *21*(6), 826–835. 10.1093/her/cyl063

AUGMENTATION | Meaning in the Cambridge English Dictionary. (n.d.). Retrieved September 13, 2020, from https://dictionary.cambridge.org/dictionary/english/augmentation

Church, T. S., Thomas, D. M., Tudor-Locke, C., Katzmarzyk, P. T., Earnest, C. P., Rodarte, R. Q., Martin, C. K., Blair, S. N., & Bouchard, C. (2011). Trends over 5 Decades in U.S. Occupation-Related Physical Activity and Their Associations with Obesity. *PLoS ONE*, *6*(5), e19657. 10.1371/journal.pone.0019657

Csikszentmihalyi, M. (2008). *Flow: The Psychology of Optimal Experience*. Harper Perennial Modern Classics.

Drinkwater. (1996). *Physical Activity and Health: A Report of the Surgeon General - Google Books*. https://books.google.co.jp/books?hl=en&lr=&id=keYhAQAAMAAJ&oi=fnd&pg=PA6&dq=.US+Department+of+Health+and+Human+Services+(1996)+Physical+Activity+and+Health:+A+Report+of+the+Surgeon+General.+Department+of+Health+and+Human+Services,+Centers+for+Disease+Cont

Duhigg, C. (2012). *The Power of Habit: Why We Do What We Do in Life and Business*. Random House.

ESP32 GREY Development Kit with 9Axis Sensor – M5Stack-Store. (n.d.). Retrieved May 30, 2020, from https://m5stack.com/collections/m5-core/products/grey-development-core

GAMIFICATION | Meaning in the Cambridge English Dictionary. (n.d.). Retrieved October 5, 2020, from https://dictionary.cambridge.org/dictionary/english/gamification

Gerling, K. M., Miller, M., Mandryk, R. L., Birk, M. V., & Smeddinck, J. D. (2014). Effects of balancing for physical abilities on player performance, experience and self-esteem in exergames. In *Proceedings of the SIGCHI Conference on Human Factors in Computing Systems*, 2201–2210. New York, NY United States: Association for Computing Machinery. 10.1145/2556288.2556963

Hefner, D., Klimmt, C., & Vorderer, P. (2007). Identification with the player character as determinant of video game enjoyment. *Entertainment Computing – ICEC 2007. ICEC 2007. Lecture Notes in Computer Science, 4740*, 39–48. Berlin, Heidelberg: Springer. 10.1007/978-3-540-74873-1_6

Kajastila, R., Holsti, L., & Hämäläinen, P. (2016). The augmented climbing wall: High-exertion proximity interaction on a wall-sized interactive surface. In *Proceedings of the 2016 CHI Conference on Human Factors in Computing Systems*, 758–769. New York, NY,United States: Association for Computing Machinery. 10.1145/2858036.2858450

Kishishita, Y., Das, S., Ramirez, A. V., Thakur, C., Tadayon, R., & Kurita, Y. (2019). Muscleblazer: Force-feedback suit for immersive experience. In *Proceedings of the 26th IEEE Conference on Virtual Reality and 3D User Interfaces, VR 2019*, 1813–1818, IEEE. 10.1109/VR.2019.8797962

Klimmt, C., Hartmann, T., & Frey, A. (2007). Effectance and Control as Determinants of Video Game Enjoyment. *Cyberpsychology and Behavior, 10*(6), 845–847. 10.1089/cpb.2007.9942

Kondrič, M., Sindik, J., Furjan-Mandić, G., & Schiefler, B. (2013). Participation Motivation and Student's Physical Activity among Sport Students in Three Countries. *Journal of Sports Science and Medicine, 12*(1), 10–18.

Mueller, F., Khot, R. A., Gerling, K., & Mandryk, R. (2016). Exertion Games. *Foundations and Trends® in Human–Computer Interaction, 10*(1), 1–86. 10.1561/11 00000041

Nojima, T., Phuong, N., Kai, T., Sato, T., & Koike, H. (2015). Augmented Dodgeball: An Approach to Designing Augmented Sports. In *Proceedings of the 6th Augmented Human International Conference*, 137–140. New York, NY, United States: Association for Computing Machinery. 10.1145/2735711.2735834

Oh, Y., & Yang, Stephen P. (2010). Defining Exergames & Exergaming. In *Proceedings of Meaningful Play 2010*, 21–23. https://meaningfulplay.msu.edu/

Pate, R. R., Macera, C. A., Pratt, M., Heath, G. W., Blair, S. N., Bouchard, C., Haskell, W. L., King, A. C., Buchner, D., Ettinger, W., Kriska, A., Leon, A. S., Marcus, B. H., Morris, J., Paffenbarger, R. S., Patrick, K., Pollock, M. L., Rippe, J. M., Sallis, J., & Wilmore, J. H. (1995). Physical Activity and Public Health: A Recommendation from the Centers for Disease Control and Prevention and the American College of Sports Medicine. *JAMA: The Journal of the American Medical Association, 273*(5), 402–407. 10.1001/jama.1995.03520290054029

Penado, F. J., & Dahn, J. R. (2005). Exercise and Well-Being: A Review of Mental and Physical Health Benefits Associated with Physical Activity. *Current Opinion in Psychiatry, 18*(2), 189–193. 10/cfcxb2

Physical activity. (n.d.). Retrieved June 1, 2020, from https://www.who.int/news-room/fact-sheets/detail/physical-activity

Pokémon GO. (n.d.). Retrieved August 4, 2020, from https://pokemongolive. com/en/

Rebane, K., Hoernmark, D., Shijo, R., Sakurai, S., Hirota, K., & Nojima, T. (2021). Developing the Thrower Detection System for Seamless Player-Ball Interaction in Augmented Dodgeball. *Journal of the Virtual Reality Society of Japan, 26*(2), 10.

Rebane, K., Inoue, Y., Hörnmark, D., Shijo, R., Sakurai, S., Hirota, K., & Nojima, T. (2021). Augmenting Team Games With a Ball to Promote Cooperative Play. *Augmented Human Research, 6*(1), 1–13. 10.1007/s41133-021-00045-3

Rebane, K., Kai, T., Endo, N., Imai, T., Nojima, T., & Yanase, Y. (2017). Insights of the augmented dodgeball game design and play test. In Proceedings of the 8th Augmented Human International Conference, 1–10. New York, NY, United States: Association for Computing Machinery. 10.1145/3041164.3041181

Ryan, R. M., & Deci, E. L. (1985). *Self-Determination Theory and the Facilitation of Intrinsic Motivation, Social Development, and Well-Being Self-Determination Theory.* Ryan.

Truffle hunting in Italy - Traditional truffle hunting and eating. (n.d.). Retrieved October 5, 2020, from https://www.beerandcroissants.com/truffle-hunting-in-italy/

Visek, A. J., Achrati, S. M., Manning, H., Mcdonnell, K., Harris, B. S., & Dipietro, L. (2015). The Fun Integration Theory: Towards Sustaining Children and Adolescents Sport Participation. *Journal of Physical Activity and Health, 12*(3), 424–433.

Vuori, I. (1998). Does Physical Activity Enhance Health? *Patient Education and Counseling, 33*(SUPPL.1). S95–S103. 10.1016/S0738-3991(98)00014-7

What Is Blind Soccer | Warriors. (n.d.). Retrieved October 4, 2020, from https://www.warriorsblindsoccer.com/history-of-blind-soccer

What is SUP Yoga? | ISLE | Blog. (n.d.). Retrieved October 4, 2020, from https://www.islesurfandsup.com/what-is-sup-yoga/

Wheelchair Basketball. (n.d.). Retrieved October 4, 2020, from https://tokyo2020. org/en/paralympics/sports/wheelchair-basketball/

Yanace on Twitter: "超人スポーツLunar G Ball. (n.d.). Retrieved October 7, 2020, from https://twitter.com/ya7ce/status/1040623982674763777

9 The Relevance of a Gamified Football/Soccer Development Platform

Kenneth Cortsen and Daniel Rascher

9.1 Introduction

The evolution of technology and data has become a prevailing strategy to catapult sporting performances and the sports business sphere forward (Cortsen & Rascher, 2018). This allows gamification aspects to influence football (soccer in the U.S.) development at both elite and recreational levels. In this context, gamification refers to the application of game mechanisms to make learning in football more fun. This is based on ideas for engagement, autonomy, story, and meaning as games add meaning to football experiences and produce motivation to improve, but at the same time allowing people to learn from trial-and-error approaches (Kapp, 2012). However, research gaps exist when striving to understand how gamified football training experiences may facilitate relevance concerning football players' learning curves. This chapter provides a qualitative and in-depth understanding of this question while investigating how Goal Station, a technology- and data-driven training system, and its integrated gamification elements, fits into contemporary football training and football development. Goal Station provides the case setting for this research although the authors acknowledge that there are many similar and differentiated interactive training systems and methodologies on the market, e.g., COPA 90[1] in the Bay Area, SAP 'Helix',[2] or the Footbonaut[3] to mention a few. (Figure 9.1 and Table 9.1).

9.2 Why Is This an Interesting and Relevant Field of Study?

The vision of Goal Station is to create a gamified, motivational, and interactive football training methodology, which applies technology and data to build on the learning curve of football players by adding levels of progression over time across various football competencies, e.g., technical, tactical, physical, and mental (Goal Station, 2021). Therefore, blending gamification and football learning is a matter of immersive engagement and problem solving through the use of playful scenarios for simulation and exploration of football (Busarello, 2016; Cairrao, 2020). This aspirational strategic direction, despite being very ambitious, is in sync with how

DOI: 10.4324/9781003205111-9

Figure 9.1 Former Danish national team player Rasmus Würtz demonstrating the
Goal Station training system at a youth football camp in Danish club
Aalborg BK (authors' photo).

technology and data have influenced other areas of society and business.
Goal Station is aligned with scientific perspectives to training and development, e.g., the deliberate practice theory (Ericsson, Krampe & Tesch-
Römer, 1993), and the effectiveness of simulations in teaching players via
(re)creating game-relevant situations.

However, this training methodology cannot stand alone given the multidimensional and complex characteristics of football in which the four dominant factors, i.e., technical, tactical, physical, and mental, are supplemented
by social, communicative, creative, and managerial and leadership competencies (Rasmussen, Rossing, Cortsen & Byrge, 2021). It may also be meaningful to include a deliberate play (Baker, Côté & Abernethy, 2003; Côtè,
Baker & Abernethy, 2003, 2007) approach to a team's overall training
methodology and team-related and game-specific tactical sessions, such as
11 v 11 or more detailed tactical sessions, e.g., the practice of defensive and
offensive playing patterns and other elements from a typical training schedule
during a football season.

The gamification element of football training is relevant because it holds
elements of deliberate play and intrinsic motivation (Côté, 1999; Ryan &

Table 9.1 Illustration of the Goal Station training system and its features and benefits (Goal Station, 2021).

Purpose behind training system	*A data-driven and interactive training system*	*Four main elements of Goal Station*
To intensely focus on training aligned with players' development objectives and the club's training and/or playing philosophy. Focus is based on deliberate practice and therefore on as many repetitions as possible to maximize the training output. The interactive and technology- and data-driven methodology produces football training gamification experiences aligned with the needs of modern-day sports consumers and young IT-savvy generations.	Goal Station provides an interactive learning and development platform for football players by giving clubs, coaches, players, parents and other stakeholders valuable training, learning and development feedback via its innovative and patented data technology. The system provides live data instantly on people's smartphones. While coaches access and monitor the training system through special designed Ipads, the Goal Station Player App delivers training, learning and development data instantly to players' smartphones. Thus, players can track their performances for every drill. They can unlock historic data and track their development over time. They can also use the app to compete with friends using the App's ranking feature and thus become more motivated to train and play. The App stores and collects all results from all Goal Stations around the world so all drills count.When a coach connects his/her coaching profile to his/her players, training	• **Physical products:** Goal Station offers a range of physical products to help coaches and players get the most out of their training sessions. Physical products include Goal Station Arena (comprehensive training system used as foundation for this case study, including Focus 360°/360° orientation, Rebounders, Focus Wall (shooting wall), Goal Targets (blanket to attach to the posts of the goal with included targets), Ball Training Machine (BTM), Air Player Dummies and Footpool (hybrid combination of football and pool). • **Training methodology:** Goal Station provides a range of game-inspired drills that help coaches and players practice what they need the most on the pitch; first touches, first-time passes and short passes. Maximize players' outputs of the training sessions by up to five-doubling the number of touches with the ball compared to a normal training session. • **Technology:** Combine Goal Station products with the Ignite Trainers and get even more out

(Continued)

Table 9.1 (Continued)

Purpose behind training system	A data-driven and interactive training system	Four main elements of Goal Station
	results are transferred instantly. This allows coaches and players to track passes and drills in terms of precision, performance over time, strengths and weaknesses while allowing them to use data to get evidence in respect to where players need to learn and develop. By the use of intelligent lights and easy-to-use software, it is possible to track every pass, shot or dribble in every drill. The advanced software is specific to Goal Station with pre-set drills and opportunities to design additional drills, which make the system simple and flexible to use, e.g., adaptable to specific training and/or playing philosophies.	of training sessions. The Ignite Trainers are placed on the Goal Station equipment to indicate which light players should pass/shoot/dribble to. The training methodology does not only improve players technical skills. It also improves their cognitive skills and their orientation. • **Data:** Track players' performance live during training sessions, use advanced data analytics to get insights about individual player performance and track development of players' performance over time. Players can also apply the Goal Station Player App to see their best results, see their training history and compete with friends.
Benefits for clubs • The commonality derived from this training methodology helps to ease the transition from youth to senior player. • Creates a baseline for comparison across age, gender and generation. Makes it possible to see improvements. • The test opportunities via the system can be designed so that testing uses the same exercise-sacross age, gender and generations where only	**Benefits for players** • Improves the players basic skills, e.g., first touches, first-time passes and short passes, which accommodate the majority of the actions of football players in game-specific contexts. • Increases motivation for individual training via gamification. • Allows the players to compare their skills with teammates.	**Learning output** • The output is a comparable database, which helps to raise the level of basic skills for the entire squad. • Allows clubs, coaches and players to compare data across age, gender and generations. • Gives opportunity to create an overview of the squad's skills. • By generating data from each and every exercise, an instant insight into players'

(*Continued*)

Table 9.1 (Continued)

Purpose behind training system	A data-driven and interactive training system	Four main elements of Goal Station
the level of difficulty changes. • There is a high level of design flexibility when it comes to specific drills applied to improve the skills desired by a club, its coaches and players. • While regular training sessions typically give players between 150–250 football actions depending on the intensity, Goal Station usually offers players 800–1,100 actions. • Take control of your training in terms of enhanced gamification via focus on the intersection between passes, ball control and repetitions and the features of a technology- and data-driven training system. • Opportunity to engage in specific, measurable, attainable, realistic and time-bound (SMART) test processes focused on specific drills, learning and development objectives.	• Maximizes the number of ball touches. • Also encompasses the junction between important technical, tactical, physical, mental, social, communicative, leadership and managerial competencies.	performance and development over time is created. • Creates a tool that can be used to track progression on a specific team or across all teams in a club. • Can also be used as a feedback tool for coaches whose age groups do not follow the clubs' norm for basic skills.

Deci, 2000; Deci & Ryan, 2007). Gamification can be defined as "the practice of making activities more like games in order to make them more interesting and enjoyable."[4] Kapp (2012, p. 7) defines a game as *a system in which players engage in an abstract challenge, defined by rules, interactivity,*

and feedback, that results in a quantifiable outcome often eliciting an emotional reaction. As such, this leads to a definition of gamification suited for this football learning context, in which *gamification is using game-based mechanics, aesthetics and game thinking to engage people, motivate action, promote learning, and solve problems"* (Kapp, 2012, p. 7). Goal Station benefits from the interactive features in the intersection between technology and data and the capacity of gamification to improve some of the above-mentioned competencies for players of all ages and levels.

9.3 Methodological Case Study Considerations

This case-study applies a pragmatic (Dewey, 1916; McDermott, 1981) approach inspired by symbolic interactionism (Mead, 1934; Blumer, 1986). This inductive study is grounded in a qualitative, explorative and interpretive research tradition because sport is a social phenomenon, and behavioral patterns in this football learning and development context cannot be comprehensively understood in causal and positivist relationships (Gratton & Jones, 2014).

Thus, time and context matters. The football setting is the Goal Station training facility located in a football club in Denmark, with teams ranging from the highest elite and professional levels to the recreational levels of football across gender differences. Data collection includes participant observations, semi-structured interviews with purposely selected respondents, i.e., one focus group interview, four e-mail interviews (due to COVID-19) and one face-to-face interview (Kvale & Brinkmann, 2009). More information on the respondents can be found in Table 9.2. The respondents were carefully selected due to their contextual understanding of Goal Station.

9.4 Data Application and Performance Data

Data helps to understand new elements of football. Football is a very complex and dynamic game with many nuances and data can (if qualified) provide meaningful contextual understanding and boost sporting and business performances in football (Cortsen & Rascher, 2018). Data can help to put new knowledge in play. Knowledge and skills produce competencies, e.g., "the sets of behavior that a person must display in order to perform the tasks or functions of a job" (Hayes, Rose-Quirie & Allinson, 2000, p. 93). In this case, data may influence competencies in relation to how players perform specific activities with observable results through application of their knowledge, attitude, and skills. For instance, this happens through passing and receiving the ball in specific contexts (drills) while measuring elements such as the total time, the average response time and the precision concerning hit rate and misses. (Figure 9.2).

Table 9.2 Overview of respondents from the interviews

Gender of focus group respondents	Occupation of focus group respondent
Respondent 1a: Male	Manager from Goal Station
Respondent 1b: Male	Men's Head Coach, professional football club that plays in the top Danish men's league and has domestic championship and FA Cup titles, including participation in UEFA Champions League and UEFA Europa League tournaments
Respondent 1c: Male	Women's Assistant Coach, professional football club that plays in the top Danish women's league
Respondent 1d: Male	Men's Reserve Team Head Coach, professional football club that plays in the top Danish men's league and has domestic championship and FA Cup titles, incl. participation in UEFA Champions League and UEFA Europa League tournaments
Respondent 1e: Male	Background as professional football player for approximately 10 years in various clubs in England. Men's U19 Assistant Coach and former Men's U15 Head Coach, professional football club that plays in the top Danish men's league and has domestic championship and FA Cup titles, incl. participation in UEFA Champions League and UEFA Europa League tournaments
Gender of e-mail interview respondents	*Occupation of e-mail interview respondents*
Respondent 2a: Male	Head of Football for the club's elite youth teams on the men's side
Respondent 2b: Male	Youth Development Coach and Head of Recruitment for the club's elite youth teams on the men's side
Respondent 2c: Female	Central midfielder for the professional women's team
Respondent 2d: Male	Central midfielder for the U19 elite team on the men's side
Gender of face-to-face interview respondent	*Occupation of face-to-face interview respondent*
Respondent 3a: Male	U10 boy's player in a partner club and participant in all the club's recreational football camp activities in the past six months

Data assists in unfolding how a player's competencies reach beyond his/her knowledge and skills and meet task completion prized in an organizational learning context, as valued by coaches and in alignment with a specific playing philosophy (Chyung, Stepich & Cox, 2006). Hence, the

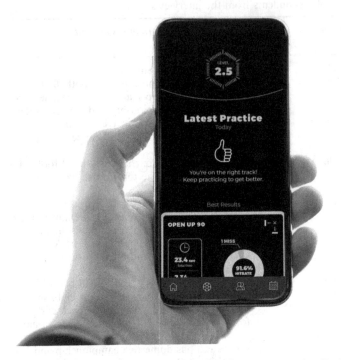

Figure 9.2 An example of a player's training data from the Goal Station App (authors' photo).

interaction between technology, data, and the competency-related organizational learning context, which is featured in Goal Station can help to build "a cluster of knowledge, skills or abilities," which "provides the means to effectively perform activities of a given job" (Fahrner & Schüttoff, 2020, p. 347).

This reflects Goal Station's multidimensional training tool functionality, but it also accentuates the importance of organizational relations and learning (Senge, 1990).[5] Respondent 1e notes that data can give indications of *How the individual player improves*. However, data must always be qualified by asking relevant contextual questions, e.g., *Would you also do this in a game?*, which is also supported by research (Cope, Partington, Cushion & Harvey, 2016) that recommends that in football coaching, players should be included in the learning process via the use of ongoing questioning.[6] The same respondent notes that *It is about getting the ball to your teammates quickly in games*. Goal Station offers reflections about the game and thus important learning points. Respondent 1e states that Goal Station may deliver relevant game-related learning via the ability of exploiting *time to scan*.[7] These notions show that data and coaching are interconnected.[8]

Respondent 1c finds some differences between coaches and players in that he *likes the moment there (ed. in Goal Station) in the training part more than the data part. I like to coach in relation to the small details in the execution* although his players *apply the following response time to look at their time and to compete [...] I got 23, what did you get?* So, there is a constant interaction between the technology, the data, sporting performances (including competencies) and the organizational learning context (including factors such as coaches, coaching, playing philosophy, playing style, and the balance between skill training and game-relevant training factors).

This study found an interesting behavioral tendency in that Respondent 1a emphasizes that *young people today are used to everything being systematized.*[9] Respondent 2a notes that *If you look at society in this relation, then yes,* but "Goal Station also gives you fine opportunities to create a competitive environment, which we should grow even more," and this cultivation is a vital opportunity with the technological and data-driven interactivity in mind. Respondent 2b adds that Goal Station is very functional as

> *a tool for the nurturing of the most basic (and most frequently used in games) skills, i.e., passing and receiving. Hopefully, it can also be developed into a learning room in which you can experiment and play. Goal Station challenges many of the uniform movements over a short time. Goal Station also challenges due to the fact that football is a running game.*

This training system is therefore a sharp contrast to *just taking a ball and then you stand in the school yard or at home in the garden* because this scenario *isn't really there anymore because there are some people, who must come and tell you that now there is time for you to stand in the school yard.* As such, initiative and drive are essential cornerstones in building positive momentum and thus activating development forces when it comes to personal football development, in alignment with research findings (Ryan & Deci, 2000; Pink, 2009). Therefore, *self-guided training is very central and significant,* whereby Respondent 1b states that *My experience tells me that they know it (ed. that self-guided training is important)* and therefore it is important that coaches catch this momentum by telling the players that *I am here if you have any doubts* because this can inspire[10] them to come and *say that I would like to train this and try this.* Data can help to give answers and measurement, which is in relevant alignment with the systematized everyday life of these young generations.

9.5 Individual Development and Self-Guided Training

Football has gained positive development from increased levels of professionalization and commercialization. This has sparked an evolution of interactive technologies and data applications, which is a contrast to how individual and self-guided football training existed decades ago. Goal Station exemplifies this as Respondent 1e highlights that it is good

> to use when you are alone or when you are two, maybe three. Then, the facility is great. The facility is already there so you can take a bag of balls or two to three balls and then you run something as self-guided practice.[11]

The system is highly interactive and delivers feedback and as Respondent 1c puts it, Goal Station *remains a competition with yourself all the time. It is about yourself and about becoming better and better. You cannot blame others.*

Answering whether individual training and self-guided training is important for the development of football players, Respondent 1b emphasized that

> It is pivotal [...] The way it is today, this has become attached. You don't have the 8 hours in the yard, which you had in the old days, so you need to find them in another way. So, if you want (ed. to accomplish) something, then you must go this way. If not, you don't have a chance!

Respondent 1e noted that in youth football, there is a tendency for clubs to prioritize to *give them some time for individual training,* but the experience is that *the boys and the girls must learn. They must learn how they should train so they must set some goals*[12] in terms of how they want to train. It may *seem like they don't do anything,* but one perspective on this is that we *have become so structured that we are not allowed to explore.*[13] The respondent stresses that

> with Goal Station, you can stand and fall in love with the ball, how to kick it so that it spins and how you... Yes, whether it is a volley or with the inside of the foot, I think that there is an element in this self-guided training, which is vital. Well, first what do you train technically, but also with the younger, they must learn what it means to engage in self-guided training, to set some technical goals for the day.[14]

Respondent 1b nods when answering the same question and says

> Yes of course [...] And the problem is not Goal Station but the accessibility[15] to it [...] I need to go get a tablet, ok there was no power on it. Then, one of the lights doesn't work. Where are the balls? Are there others on the pitch? It is this environment. Well, for me it is indisputable that it is good. This is not the problem. It is the environment. It is the everyday life. This is what we need to discuss. All the other stuff is indisputable [...] How can I go and use it whenever I want? How can Kurt? How can Anders? [...] Otherwise it holds no value.[16]

Concluding on this question, individual or self-guiding training is good for football development and interactive technologies can support this development.

There is no doubt that interactive technologies, data, and Goal Station form an interesting training solution, which underscores football development, but it becomes a matter of accessibility, discussed below.

9.6 Goal Station as an Applicable Training Setup

As proposed above, Goal Station is a relevant training system in modern-day football development. One of the strengths of this interactive and technology- and data-driven training system is its multidimensional features. Yet, one of the weaknesses of Goal Station may be accessibility.[17] Respondent 2c mentions that *Goal Station makes it easy to train alone because all the drills have been made for one or a few players at the time and in addition it is easy because the drills have already been made in advance.* However, she addresses that

> *I need information about how to do it myself when it comes to where and when you can use the system and the lights and the Ipad as well as how you set it all up. These things could be more accessible.*

This emphasizes that clubs, which have invested in the training system, should consider player access to optimize learning because the system is a great supplement to the learning curves in football clubs' talent development infrastructure. Respondent 2a states that the methodology means *A general overload in terms of repetitions and massive transfer/motor skill training of the body.* Therefore, it is *An absolute necessity […] I expect that it will have a very positive impact.* Respondent 2b finds the system *Decisive. It is merely another information which is a foundation for all the decisions we make and so an upgrading of the qualifications for decision making,* so it can also play a role in the *Selection, prioritization of training content and motivation,* and *As something which gives the players an opportunity to play and to nurture.*

9.6.1 Multidimensional Training System

The data reveals the premise that Goal Station is a multidimensional training setup and Respondent 1e mentions the meaning of its varied nature, e.g., *I think that there are many things. There is not only one thing, which is most important,* while Respondent 1d states that, *Basically, you can use it for many things and it is relative easy to design it situationally in relation to games.* The same respondent elaborates on the progression of player development as one essential feature as *You can develop it so that you can go from one basis drill to the ability of simply and easily adding more and more layers.* Thus, users can start with *some basis drills, some simple passes, and touches but you can also add shooting.*

Respondents recognize that data application must include a varied and multidimensional approach. Respondent 1d says *Data is never good if it only depicts a snapshot. Data must be applied over time.* Consequently, the multidimensional element comes into play in that coaches and players should focus on *to move the good time to something on a football pitch.* He continues by saying that *If data is only used for a funny competition on a Monday afternoon,* then *It is irrelevant.* However, if coaches and players are capable of attaching the transferability to *take it into a full-size pitch, then it is a very good tool.*

The transferability of learning and decision-making to a game-related context is valuable to any talent development environment in football, and Respondent 1a argues that *Sporadic data are never relevant.* He stresses that it is only through *continuous data over a long period where you can see some kind of development […] The single training and the single data point are not the relevant part. As you know, it is the great picture.* To sum up, there are some fundamental ways in which the interactions via Goal Station facilitate an enhancement of the learning curve in football development. Subsequently, clubs and coaches should embrace this when implementing such interactive training systems whether the training takes place in a team- or an individual-oriented setting but at the same time be cognizant about the fact that data should be qualified for contextual purposes (Cortsen & Rascher, 2018).

9.6.2 The Framework of Goal Station and the Bridge to Autonomy, Mastery, and Purpose

Regarding the learning curve and outcome, Respondent 1e highlights that *it is good for your basic skills to help you get more surplus in the bigger play. So, it is a part of many things, which are important as a football player.*[18] Goal Station is designed in an interactive way, which can orchestrate simulated game-relevant "football situations." Moreover, the system provides more of these situations and repetitions than players usually experience in ordinary team-based training drills, e.g., 4 v 4, 7 v 7, or a full-size 11 v 11 playing session, in which there is often much tactical focus. In contrast, Goal Station is designed, according to Respondent 1a, so that *You must do things over and over and over again […] Goal Station is a unique offering and creates a really unique frame for going through these specific repetitions.* Repetitions and situated learning can both influence football development positively as there are ongoing interactions between learning and the social context in which it takes place (Lave & Wenger, 1991; Gréhaigne, Griffin & Richard, 2005). Research (Hung & Chen, 2001, 2002; Lunce, 2006) provides evidence that simulations act as a method in which contextual and game-realistic or situated learning may be facilitated. The comfort zone model illustrated in Figure 9.3 is often applied in educational settings to inspire players to "stretch themselves" to move out of their comfort zone and thereby go beyond their defined constraints and by inference learn. This learning approach is based on cognitive development (Piaget, 1977) and dissonance (Festinger, 1957), and this football learning

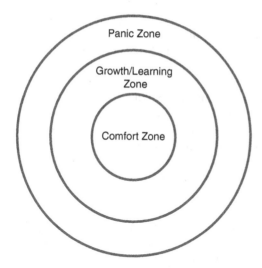

Figure 9.3 The comfort zone model applied for football learning (inspired by Brown, 2008).

happens in the growth and learning zone, but without being negatively influenced by entering a panic zone where the gap between current reality and desired future capabilities is too wide and thus provokes feelings that will block the constructive process of learning (Brown, 2008).

In a response to whether it is effective to present players with examples of development without having concrete evidence, e.g., data, Respondent 1b admits that *It is important to have evidence and data may be very beneficial in this regard.* Respondent 1e adds that *Things become automatic for you* and if you do things *with the right technique, which coaches help you with, then it becomes automatized* and it turns into applicable *muscle memory.* Yet, he also declares that *it is complex* and that *Everything takes place in context [...] How was your last action? Are you tired?* He proposes the importance of repeating things in that

> *It is this automatization as if you play the guitar or the piano, practice something repeatedly so that it at some point sits in your body and then it frees some capacity in the brain so that you can make the right decision.*[19]

Thereby, Goal Station is aligned with how Daniel Pink (2009) addresses the importance of autonomy, purpose, and mastery as criteria, which help to motivate, inform, and improve training practices. However, clubs, coaches, and players must apply critical thinking when working with interactive training systems such as Goal Station as Pink (2009, p. 56) underscores this: *Take mastery. The objective itself is inherently long-term because complete mastery, in a sense, is unattainable. Even Roger Federer, for instance, will never fully* master*the game of tennis.* For instance, superstars like Federer are typically never fully satisfied and can often come up with a list of things to improve after their performances. It is important to stress that just like the functioning of computers are dependent on human interaction and the application of competencies, i.e., contextual understanding, the same goes for Goal Station as the training system and the data cannot stand alone. Pink's thoughts about mastery illustrate that *introducing an* if-then *reward to help develop mastery usually backfires. That's why school-children who are paid to solve problems typically choose easier problems and therefore learn less (Benabou & Tirole, 2003). The short-term prize crowds out the long-term learning.*

In the focus group, there is a discussion about the role of players versus the role of coaches and Respondent 1c perceives that

> *I just think that Goal Station data are more valuable for the players than for the coach [...] It is the data that they receive on their phones (ed. via the Goal Station App), which they can use to move themselves because it is so situational.*

He explains that *It is tough for us to see all situations* so what the players receive after the drill may be *more valuable than me going through the development of 26 players because there are so many things taking place* because *the individual response to the player* is significant and also plays a role in *maintaining the competitive element.* However, when unfolding this discussion, Respondent 1a, who possesses a cultivated data background, points that there is value in coaches *drawing the data of each of the 26 players up in a form of roster context.* This allows the coach to study

> *the entire roster with an interval range, which may be applied to give the coach an overview as he/she gets a range instead of an exact average [...] If the players are within this range, it matches somehow with the skills, which we anticipate that they have [...] Those, who are slower, will be found, and you can work on getting them into the range.*

In saying this in the context of a six-week testing period for the best elite teams on the men's side in the club, the same respondent notes that *most of the rosters have in fact moved in terms of the drills toward getting all the players into the range.*

9.7 Societal Changes, Communication and Facilitation of Football Development

The empirical data reveal that Goal Station and similar systems have the potential to become well-positioned tools in the future of football training. Respondent 1b remarks that he views it *Of course as a part of it. It is just a matter of accessibility* while Respondent 1c remarks *Totally agree* but there is a need to make it easier for the players because right now *it is dependent on us (ed. coaches)*. One way to solve this could be that the club *creates some intervals every day where you have some people employed, who say that from 5 to 8 p.m. today there is training in Goal Station and the players just have to show up.* Respondent 1a says *I agree that the accessibility is vital* while Respondent 1e comments that *I see Goal Station as a supplement to basic training,* but there is a need to make sure that the football coach doesn't *opt out when it comes to the talk about transferability.* Accordingly, it is important to understand that while Goal Station and similar systems hold many basic training benefits (e.g., technical skills, repetitions, concentrated physical actions with the ball etc.), gamification aspects and a modern-day approach to data-driven football training, these systems also do not *substitute a 4 v 4* drill as *4 v 4 is much more complex.*

However, Goal Station offers, according to Respondent 1e, an opportunity to *train the basics* in which players *train some forms of repetitions* which they *have to go through* and here the system *can add something in this element, which is very important.* Respondent 1b takes this a step deeper and observes that you can benefit from concentrated actions as *it holds immersion* and *when I say to Player X[20] that now we are going to train turns,* then it will provide *absorption in this situation so it has a huge value.* Respondent 1a grasps that it is vital that Goal Station *comes in and takes the right place in a club and that it isn't perceived as something that will take over but rather as a supplement* so that the club doesn't send the message that it *sets up a fitness room and then says that now we must train conditioning and strengthening and by the way it will substitute eight tactical training sessions.* Instead *it must support the current training, the immersion and the focus on repetitions, on the basic training of the skills, which you would like the players to adopt so that they hit the mark* because at the end of the day *the players that repeat their basic drills and excel in this regard they have a bigger chance to succeed.* The latter is a result of the fact that *to automatize more and more will pay off. This is how it is with all technology. It has to start a place and as soon as you think that it is applicable for specific areas, then you demand more automatization.*

As Respondent 1e proposes, the communication elements in Goal Station hold crucial meaning, cf.

> We have run something with communication. When you are alone, then you can do passes but you don't really practice communication [...] You can create a drill where one player passes to the rebounder and then he/she must go to another rebounder. However, we need a combination (ed. with other players) before we move on.[21]

Communication is a vital ingredient in any team sport and to facilitate this in football development, Goal Station integrates this aspect. The respondent details this in that

How many times haven't we stood there as coaches and screamed 'come on, we need to talk more'! But how do you actually train this? Therefore, this is very specific that we cannot do the drill unless we communicate. This is perhaps a good layer, which you can add [...] Maybe, we cannot measure it as well (ed. with data) but this is not that important. Well, if we train this, then we won't become worse.

Still, Respondent 1c regards that it is essential that the club *educates all players to be able to initiate it, to do the drills themselves. And this is a big bite even for us, who use this a lot* while Respondent 1b responds *Yes, there ought to be 2–3 tracks, which are running all the time.* This is backed by research (Werner & Dickson, 2018), which concludes that knowledge sharing has a positive influence on elite players' development and performance, while it presents the need for future knowledge management tactics to exploit this untapped potential, which Goal Station may facilitate when utilized.

9.7.1 *Influence of Coaches*

In a comment about whether individual and self-guided training is important, Respondent 1d mentions *I agree,* but highlights that the other coach worked *with a good group, which most likely knows what to train [...] A coach's or manager's task is maybe to guide them regarding their self-guided training and why we need the individual training.* He suggests that there is a positional element to this in that

If I have a right back, who continues to do self-guided training and kick balls to the goal, how much does he/she gain compared to going to the pitch and finding him-/herself in situations where he/she practices when he/she needs as a right back such as touches, passes, positioning etc. [...] the most important task is to guide them in the right direction.[22]

However, the deliberate play and creative approach, discussed below, may promote relevant learning,[23] but there is definitely a need to enable collective goal fulfilment. This brought up a discussion of how Goal Station is dependent on the interaction with coaches to foster an optimal learning curve. Respondent 1b emphasizes it to be *very important to have guidance, which helps* instead of the fact that *individual training only took place if the coach was there* because then it is not individual training, but only the coach's team. Thus, it should be emphasized that it is all about *the player working with him/herself.* While acknowledging that *We are only resources as coaches,* Respondent 1b refers to the strong foundation that lies in *They fall in love with football and it doesn't happen if we (ed. Coaches) control things all the time.* He goes on to note that *Then, they will fall in love with us.* However, this is not necessarily the case when it comes to interactions between coaches, players, and a very systematized and formalized football-world in which there is a need for a balance with "deliberate play" to be able to develop and retain people in the football environment.

9.8 Deliberate Practice and Expertise in Sports

As research highlights, *individuals begin in their childhood a regimen of effortful activities (deliberate practice) designed to optimize performance* (Ericsson, Krampe & Tesch-Römer, 1993, p. 363). Therefore, analysis of expert performance is related to opportunities and constraints regarding contextual adaptation and learning. In addition, it concludes that escalations in duration, intensity, and organization of training hold relevance in talent development.

9.8.1 Intensity of Play and Time Optimization

Concerning the design of football training, Respondent 1c notices that *I have trained elite women but also young kids* and in these sessions it has been a huge help that Goal Station *is framed and that the ball never flies away.* As such, *You are always in motion, and this is definitely a huge advantage when you coach young kids that the balls aren't kicked away.* Thereby, teams can *have focus on football and on developing the technical skills.*

Respondent 1a emphasizes the effectiveness of training time and intensity of the training experience in that *When you have to develop players, regardless of the sport, it is about repetitions, and repetitions, and repetitions,* which is very much aligned with Ericsson's (1998) research focused on developing expertise. He explains that

> *It is pretty simple, but it doesn't do well if you are at a football pitch and the ball goes over the goal and you must go and get it 30 meters away and then back again to do it again.*

Effectiveness is a function of the quality of training and vice versa as the same respondent perceives, *In relation to touches and repetitions, I think that it provides the opportunity to do a very specific repetition or a very specific touch, which may practically be impossible to train.* Conversely, research (Trowbridge & Cason, 1932; Ericsson, Krampe & Tesch-Römer, 1993) advises that when there is no suitable feedback, competent football learning and development is impossible or at least reduced even for the most motivated players. This also insinuates that isolated repetitions of a specific football task will not automatically result in positive football development. The argument is backed by how the interactive characteristic of football, and thus the influence of coaches and the facilitation of football development, are essential elements in football development. Nevertheless, the detail-oriented training opportunities provided by Goal Station are espoused by Respondent 1a, who adds that Goal Station can help players to improve details such as

The instep kick for instance, where you need to hit the ball very precisely and then do the same motion again and again without there is another one (ed. person), who spoils it. Therefore, you can refine your technique in some very specific areas, indeed in all areas.[24]

9.8.2 The Effectiveness of Training

Although Respondent 1d refers to the meaning that in Goal Station, *There are many repetitions. It is extremely important to get all these repetitions into it,* he still holds the opinion of importance that players interact with coaches to receive feedback to improve (Ericsson, Krampe & Tesch-Römer, 1993). This can be done by considering elements such as benchlearning (De Bosscher, 2015; Arias, 2019) and individual-oriented diagnosis of mistakes. Then, players, according to Respondent 1a, can find *optimization from the fact that you can do things again and continue to do it. In that regard, I think that to develop these skills in the best possible manner, it requires repetitions, and this plays well with Goal Station.*

This opinion sparked a relevant discussion between two of the respondents in the focus group session regarding the quality of training. Respondent 1a argues that *We must look at the effectiveness* while Respondent 1b comments *Yes exactly and the amount of training. Do we have to train more*[25] *and do we have to do that with more freedom? [...] But this is why it becomes more systematized.* Yet, football development practice indicates that there are constant interactions between habits (e.g., training), behavior, and culture in which good and effective football development can be found (Cortsen, Hird & Kvistgaard, 2020). This is evident in the discussion as Respondent 1a states that *If what we want to cultivate are these things then we have to have some tools, e.g., Goal Station, which Respondent 1b agreed with.* Respondent 1a completed the discussion by concluding that clubs must *support this effectiveness,*[26] *which ought to be there in order for you to break through.*

9.8.3 Technical Skills and Game Relevance

Respondent 1b regards that *the technique is the most important.*[27] He mentions that if you are part of *U19 (ed. the elite U19 team), there we have to get ready for the Super League. So, this is always a balance.* However, in a discussion of the timing related to training technical skills, he says, *I think that it is all the time* and you cannot say *Ok, you are U15, now it doesn't make sense.* Respondent 1e agrees and says *Of course not [...] When you become older, it is the small percentages, which make a difference and in this regard it is essential to have some basic memory and muscle memory.*

In matching technical training with technology, data, and measurement, it may be beneficial, but there may also be pitfalls. Respondent 1b notices that

I have seen some drills there and the speed may go up but then we often see that the technique becomes worse.[28] He elaborates in that *It is not about many repetitions. It is about the right repetitions [...] Well, time is exciting when you have the right repetitions.* Nonetheless, he exemplifies the potential pitfalls by noting that

> *I can see five players do a drill in five different ways. One player takes it with the outside of the foot, the other stops it with the foot and so on. Well, what is the right way and what is it that we want to see. In this sense, I think we must be sharper.*

In a question about the transferability in play, he acknowledges that *It is the most important* element in play.[29]

9.9 Deliberate Play and Creativity

While there is evidence in team sports that deliberate practice and time invested in a sport help to build expertise (Ericsson, Krampe & Tesch-Römer, 1993), additional research concludes that there is no single success formula or roadmap to sports expertise. Instead, this research recommends that both "deliberate practice" and "deliberate play" may support the path to expertise (Côté, 1999; Côté & Hay, 2002; Berry, Abernethy & Côté, 2008). Deliberate practice encapsulates a vastly organized training with the intent to enhance performance in the domain of specialization. Deliberate play typically takes place during the early years of sports activity, e.g., ages 6–13 years, earlier than specialization, e.g., approximately 13–16 years, and investment, e.g., approximately 17+ years, and compresses sports activities, which are intrinsically appealing, bring instant enjoyment, and are concretely organized to boost gratification (Berry, Abernethy & Côté, 2008). In other words, deliberate play seeks a more playful and creative aspect to sports participation and development, which proves imperative for motivational, recruiting, development, and retention purposes in this case and football club context where there is already an overload of deliberate practice activities.

9.9.1 Players' Motivation

Players' motivation is supported by the empirical data. Respondent 1a identifies the importance *That it (ed. Goal Station) is fun and motivating at the same time and creates a frame in which the ball isn't lost is just positive*[30] *[...] The play element is vital [...] There is a central concept there.* Respondent 2c says that

> *It is different and fun and it is a great supplement to the ordinary team training. It gives you some technical skills where you intensively get the opportunity to work with your first touches and passes and where there are many opportunities to work with both legs.*

She adds that comparing results with your teammates *creates a competitive element where you compete against your teammates' results and this helps to secure development.* Respondent 1a goes on by observing that

> *It may have been forgotten to discuss from a societal viewpoint*[31] *that you have a ball in your hand, that you are with your day in the sports centre and you sit there and do your homework or you play with the ball or walk around and dribble with a football, a basketball, or a handball.*[32]

Respondent 1b touches upon the balance between deliberate practice and time in contrast to the playful deliberate play element, e.g.,

> *You (ed. U19) train four times per week, right? [...] There may be 20 minutes warm-up so you have five hours of football (ed. training) in a week and then you have games where you have been selected [...] In the old days, you used this (ed. amount of hours) in one day so we cannot even measure this in hours how much these players lack behind*[33]

in commenting on why people with great access to facilities may have a significant benefit.

While acknowledging that data *has great relevance,* Respondent 1a states that *If you have the players and they do a drill and they do it in each player's individual way. There is free play. After this, you go in and apply coaching.* He emphasizes that it is meaningful to be able to *follow when they reach their fastest time and surpass this in the correct way [...] This is why data plays a very important role in following when and how they develop and if they can improve things technically.* Respondent 1b applies a critical perspective and underscores that Goal Station *only measures when you hit the rebounders* so he views that it means *two factors, e.g., a hard pass or it may be a good touch so you may be able to go fast without doing it right*[34] while Respondent 1a highlights that *data plays a very critical role in following if the player reaches the (ed. development) points, which we would like him to reach.* Respondent 1b agrees but mentions that *This is exactly what I say. However, we must not mix pears and bananas.*

If coaches want their players to possess specific skills, then Respondent 1a notes that *You can transfer this to some specific drills, which you continue to repeat and then you can follow this development.* Respondent 1b responds by *I fully agree [...] You can definitely take the 'locked' skills into the big game (ed. full-size). And this is what it is all about,* while Respondent 1a concludes that data is good *because you have it over such a long time, then the small parts will be leveled and then you will get a picture of how things really are.* Respondent 1b remarks that *I am a great supporter of this. I am just saying that everything must be at its shelf, so it is a part of the big game,* but it is difficult to qualify things and to transfer these things to a *football context.*

9.9.2 Competition and Gamification Element

Respondent 1e comprehends that *The more fun it is, the more engagement you will get from the participants so there is nothing new in this.* Nevertheless, the relevance is positively reinforced at a time in which society and younger generations have really been hit by the technology, IT and data wave, which has brought gaming, the use of computers, and the interconnected sense of the world into new spheres of gamification applications in areas such as education, CRM-related marketing activities for general businesses, and specifically in many parts of the experience economy, e.g., tourism and sports. Football faces the same gamification demand as the sport reflects the surrounding society and markets and holds a central role for change, growth, and innovation due to its position as the most popular sport in the world (Giulianotti, 1999; Kuper & Szymanski, 2012). Respondent 1e confirms this situation in that *You can say that football competes a lot with computer games* and given the fact that football also faces competition from other sports and leisure time activities

> *It is decisive that it is fun to be a part of [...] One way to make things fun is by turning it into a competition [...] Data makes it easy to measure and this is easy to turn into a competition.*

Goal Station holds these opportunities, which adds sporting relevance and marketability in that the system makes *it clear for you after the training how you have performed over time [...] You can see your curves for your training results* and this is a persuasive way *to turn people hooked because they think it is fun.*
   Respondent 1c clarifies that

> *We also live in a time where kids and young people are bombarded with impressions [...] If I were to stand there and make 16 passes to two rebounders without lights and the following data, then it becomes boring after five minutes [...] Lights and the fact that there is data makes a huge impression on the kids and adults, who play and this is decisive.*

Basically, he implies that the experience is turned into *something digital and living. This is crucial.* Respondent 2c remarks that the ability to see the data and the times in relation to the drills and one's individual development assists *in that you continue to develop because you don't even compete against your teammates but also against your own time, which you always can improve.* Respondent 2d adds that *It is super to come in after training where you are rewarded if you have done well in the form of high scores or results. Well, the better you have done, the more points you will be awarded* and Goal Station *fits well in this regard because you right after training will receive your results.* Respondent 2c discusses this and notes the importance of data as a motivating gamification element in that *We have not had much focus on collecting our results at our team and therefore it has been difficult to see our development. However, it would definitely create motivation for me if I could see a development in my results.*

How Goal Station can *create diversity*[35] *in the game because everyone can be a part of it* adds another pragmatic element of football gamification according to Respondent 1a. He rationalizes that *Gamification can open for making data and measurement into a tool, which can take some of them, who would normally sit at home in front of the computer, and get their kick via data, in.* He comments that *Now, they can go to the football pitch and get this here* because Goal Station can provide them with the opportunity *to see their results in a measurable way* and suddenly the competition aspect *becomes transferable to the physical and the real world but they simultaneously get their performances on the phone.* In this way, Goal Station holds the opportunity to *expand the segment of football players so that no one are lost because football only has room for those who are good enough for the youth national teams.* Therefore, the respondent stresses that Goal Station can *open for a bunch of players, who may not necessarily become really good, but they still get their kick. In this way, the amount of football players will grow.*

While Respondent 1e thinks that *it is good to run this with lights and data,* he also proposes that there may be benefits of simultaneously running *fields where there is only focus on the technique and not time. Then you can get the right technique and transfer this to the big game.* This addresses that learning is subjective and that there are other ways to learn in football than just applying methods related to technology and data. Still, he details this focus by recognizing that

> *the lights are appealing so this is something that can get people into this. Maybe not the best (ed. they are already motivated), but there must also be a good environment [...] So if the lights retain the next best and make them better then it also gives a good effect.*

Respondent 3a has participated in all the club's youth camps in the past six months in which Goal Station plays a central role and he thinks *it is fun* because *it is difficult* and it provides some *first time drills (ed. in terms of hitting the ball first time) in which you must use both legs.* Additionally, he finds it motivating that he can train with his friends and teammates and compare results, cf.

> *It is great because you can hear who was best and have the best time [...] It is fun to know that you have the best time [...] It is great because you can see if you become better or not better [...] To begin with, I didn't have as good times or only slightly good times, but I think that they have become better the last time during the camp. There, I got the best time three times.*

He notes that *It means that I train a lot*[36] *[...] I like to win, and I would like the best time every time* as he expresses that it is motivating, cf. *It is good [...] My motivation*. He says "Yes" as an answer to whether or not he can use Goal Station to indicate where he can improve his skills and he elaborates by saying that

> *It may be the case that you are not that good with your weak leg and then you have all the good (ed. passes) with your good leg and then you get 30 seconds but then it may be the fact that when you have also trained your weak leg then you may only use 15 seconds because both legs are equally good.*

He thinks that the gamification aspects in terms of competitions are *terrific* because

> *I like competitions [...] It is easier to make competitions in Goal Station because all things are competitions [...] I think that Goal Station is fun [...] You learn a lot from individual training [...] There are rebounders, lights and such. It is more individual.*

However, he also proposes that there may be kids, who don't think it is fun because *they don't like competitions and they don't like to be last. You have to be able to put up with this if you want to do this Goal Station.* This is supported by Respondent 2c, who says that *Competitions in the training create an increased focus because you always want to win. It is often more fun when there is something to play for.* Respondent 2d says that

> *Goal Stations is for everyone, large and small, so there would be a large group of potential users of Goal Station. Given that football has gone through much development in terms of technique and speed in the game, then Goal Station is a great match [...] The greatest popularity will come through the competitive part of Goal Station as I have the perception that it is essential for the performance, which is put into training.*

9.10 Game Intelligence, Group Tactics, and Football-Related Competencies

Respondent 1a points to Goal Station's progression of learning in that *Then, there is the of course the group tactical (ed. element) [...] But the technique is important to get your ideas out (ed. as a player)* and in such a capacity *It is definitely a tool* through which it can assist in building a bridge between technical and tactical elements and hence help players demonstrate game-relevant competencies. A famous Wayne Gretzsky quote reads: *I skate to where the puck is going to be, not where it has been.* Respondent 1c sees that Goal Station can cultivate the same kind of game intelligence in that

> *We have talked about the same regarding the small kids […] If they work together two and two. If one of them is in the right side and must kick the ball to the rebounder, then the other player may be able to predict the pattern of the game and move in the direction where the ball will go. So, there is something verbal but also something about relating to the ball and the game and where the ball may go.*

Respondent 1d states that *You can add situational change of direction in relation to touches […] In a relatively simple way, you can build it to be more and more game relevant and therefore create the development, which is needed.* Respondent 2a sees that the training methodology supports *Everything in terms of basic technical skills and specifically the kicking technical competencies. Bodily/movements/rhythmic […] orientation, anticipation and perception.*

9.10.1 Scanning and Improving the Vision of Players

Orientation and decision-making skills on the pitch are pivotal when assessing the quality of footballers. What researchers (McGuckian, Cole, Jordet, Chalkley, & Pepping, 2018) call visual exploratory action, i.e., scanning movements delivered via right and left rotation of the head, permits perception of a surrounding environment and assists upcoming decisions and actions. Hence football is very dynamic, so the degree to which this is an advantage for the player's following performance with the ball is probably influenced by how and when this happens. Yet, there is a supporting link between this ability and on-pitch football performances in association with higher frequencies of exploratory head movement before receiving the ball. Respondent 1e refers to the fact that *Orientation is also important in terms of where you are on the pitch* and *The better image you have before the ball arrives, the more capacity you have and the greater likelihood of making the right decisions.* Goal Station can help to facilitate this

> *If you train your basic skills such as simple passes and receptions, which Goal Station helps to do with a lot of repetitions, then these basic skills lead to you having more time because when you receive the ball, then you don't have to look at where the ball is, or if it jumps away from you so you have more time to look around and get a better vision.*

Respondent 1c adds that

> *We have talked much about the same […] When the ball is on its way back, then you can use the process of elimination. That is, look in one direction and if the light doesn't blink then I know that must go in the other*

direction. So, it is somehow the same in relation to split-vision [...] I really like what Goal Station has with a blue and a red light so that you know where to go next. That the red becomes blue.

Respondent 1e observes that

In this 360 (ed. 360 degrees drill), where you pass the ball to the rebounder and then the ball comes back quickly, depending on how close you are (ed. to the rebounder) and how hard you pass, we have talked about if you can scan [...] You try this, where the ball is on its way, then we know that it must go somewhere but I am trying to orientate myself [...] You try to see both parts. You try to look for the blue and the red light [...] This comes after you learn to turn your head because this is the first. This is split-vision [...] Here, you can specify it (ed. compared to an ordinary possession game.[37]

Respondent 1d complements these notions in that *It is something technical but also split-vision* while Respondent 1e also speaks about the progression of learning and the meaning of time and space as

I think that the first you do, is, that as soon as you pass, then you look because then you have time before the ball comes back to you again, and then you must turn your head but when people become good at it, then you can add additional layers.[38]

Football research addresses why it is important that coaches and players are consciously aware of combining technical, mental, and physical skills with game intelligence and split-vision. Through the ability to scan and thereby benefiting from orientation and combining this with tactical knowledge, the players can unfold improved competencies in relation to ball reception at a high-quality level (Cortsen, 2010).

9.10.2 Interactive Technologies and Data Are Here to Stay and Will Overcome Resistance

Even though football on a global scale has gone through new lengths of professionalization and commercialization, and has become more sophisticated in its attempt to engage with stakeholders of the game, and its detail-oriented focus toward finding a competitive edge via technology and data, there is still a need to crack the code in terms of more in-depth ways of matching data with playing style. In commenting on this, Respondent 1b finds the application of data vulnerable because *there are not two games that are identical.* He admits that the data[39] side of Goal Station is *good here because you have the same situations again and again,* but he compares this with full-size pitch games in that

It doesn't make sense because every game has its own history, and every player has had his own history up to this game. Then there is weather, then there is wind, the result, red card, yellow card. Who is in shape, who is not in shape? Where does he play? Who is he playing with? What must the opposing team try to reach? And I can continue.

These are interesting points, but as football has evolved over time and gone from association-based amateur clubs toward more of a business orientation, there are increased incentives to apply data and to measure relevant elements (Cortsen & Rascher, 2018). When we can measure some of above-mentioned elements, data helps to give us a contextual understanding, e.g., that we should not play a particular player, that we have to pay attention to the wind, or that we must be precautious and not play a particular player for 90 minutes today because there is a risk of injury which could be very expensive to the club and the player (there is also a responsibility toward protecting the players) in sporting and economic terms. Respondent 1b notes that *When talking about data collection, it must be the same circumstances every single time [...] This is a matter of figuring out what we can use the data for [...] To take some elements out of the game.* The latter may refer to data-driven concepts such as "expected goals" (Spearman, 2018) but it also only gives football leaders, managers, coaches, and players some degree of objective truth due to some of the arguments presented above.

9.11 Concluding Discussion

Through the use of a case study utilizing qualitative methods and a specific training system called Goal Station, this research explored how gamification taps into contemporary football training and talent development environments.

The study shows that there is growth potential in both gamification capabilities and applications in football as technology use and data-driven training approaches have increased. Moreover, this development demonstrates that although gamification and interactive technology- and data-driven training systems supplement rather than substitute traditional approaches to football training, this training methodology can incentivize learning and thus positively reinforce football development. This happens across all central factors as this methodology may help to improve players' (1) technical skills through increased repetitions in reference to for instance first-time touches or precision and speed of passing, (2) positional understanding, for instance through training skills related to specific tactical positions or movements or including 360° orientation while adding the relations to teammates (involving more players in the same drill) and inserting dummies (simulating opponents) in the drill, (3) mental competencies through cognitive learning via concentrated actions of handling the ball while managing to do so with precision under time pressure and to maintain a successful performance rated over the course of for instance 12 or 16 repetitions, and (4) physical condition through rigorous drills with many relevant and context-specific repetitions.

At the same time, this methodology gives coaches and players a relevant and gamified opportunity to design immersive training situations, which can provide social, communicative, creative, managerial, and leadership competencies. It is important for coaches and players to consider these elements as the programming related to the random sequence of lights in each drill requires players to work together or to scan the environment before receiving the ball, which trains orientation and cognition. Through fun and motivating elements of competition and given the fact that the system is designed to challenge players across various competencies and quality levels (the system has eight different levels and can therefore be designed to match the needs or beginners as well as professional players), participants can enjoy measuring their improvement, which in return has a positive influence on the appeal and use of the training system and on learning and development.

Feedback exists in any qualified learning environment. As such, this training methodology supports the feedback from training in the form of tangible data that can be integrated by coaches and players into further training and improvement to account for the importance of the progression of learning. At the same time, the training methodology allows for tracking improvement and motivating players and also enhances coach and player decision-making as the system can be designed to fit playing decisions based on factors such as playing philosophy, expected style of play, and individual as well as collective learning goals.

Conclusively, repetition and situated learning are important for learning football competencies and hence also for the transferability of Goal Station to other elements of the football game. So, gamified training methodologies such as Goal Station allow coaches and players to track football learning and development over time, and offer answers to questions such as – do players improve their speed without neglecting precision? do they approve their physical condition? are they motivated to train and are there positive transfers to the full-size football aspects? The flexibility of the training methodology can be designed to fit a high variety of coaching needs while it also provides individual training opportunities without the need of a coach or the rest of the team, which can increase the amount of learning across players. The flexible, adaptable, situational, and progressive learning (or training) system is somewhat similar to a flight simulator for learning to fly a plane in that over time more and more realistic elements are added to the flight simulator to make it as relevant as can be. However, to be frank, football is more complicated than flying a plane. In fact, there are many more interactive elements with the 21 other players on the pitch (it's not as if the passengers are also trying to fly the plane too, causing complexity for the pilot). Nonetheless, this training methodology also emphasizes that the best learners in football often challenge themselves by moving out of their comfort zones and by trying new things, e.g., training the ability to play first-time passes or to take touches with the "weakest" foot. Finally, the training system is a relevant modern-day recontextualization of football learning and development with the application of interactive technology and data and gamification. Data measurement seems like an answer to the systematized life of young generations, e.g., generation Z and younger generations. However, this up-to-date approach may cause some divide in clubs between traditionalists and innovative coaches whereas the players seem to enjoy this supplementary training and learning opportunity, which is also drawing on notions of FOMO (fear of missing out) due to enormous focus on technology and data in football and in the surrounding societies. When something is at stake, stakeholders are usually alert and aware, which is the first phase in the process toward action, i.e., AIDA-processes (attention, interest, desire, action), and this places importance on the cultivation of the football culture. The latter refers to putting some thorough sporting and business strategies behind the use of the system and the methodology as such interactive training solutions possess the power to enhance football learning and development but also to boost the financial side of the clubs when catering to the needs and desires of younger generations. This also speaks toward elite football development in which it is necessary to cultivate relevant player competencies by offering learning points to develop players with better impact on the professional game and thus with high player valuations. Yet, this should account for the strategic considerations regarding the constant interactions between the technology, data, sporting performances, and the organizational learning context, i.e., including cultural fit with HRM-strategies, playing philosophy, playing style, coaching style, and the junction between skills training and game-relevant training. To sum up, there are central interactions of this system, which break down the barriers between digital games and physical football training and boost the learning curve in football development, e.g., concentrated actions and repetitions, but these interactions must always be contextually qualified whether they happen in a team- or individual-oriented setting.

Notes

1 https://copastc.com/.
2 https://blogs.sap.com/2015/11/16/soccer-goes-digital-the-helix-revolutionizes-cognitive-training-at-tsg-1899-hoffenheim/
3 https://www.bundesliga.com/en/news/Bundesliga/borussia-dortmund-and-hoffenheim-use-footbonaut-to-hone-their-passing-skills-464313.jsp.
4 According to the Cambridge Dictionary: https://dictionary.cambridge.org/dictionary/english/gamification.
5 Football is a relational and complex game; thus, the organizational relations and the interactive features of this training system is valuable.
6 Questioning activates players' cognitive processes and by considering and reflecting on the questions, they learn and become more game intelligent.
7 Check later section about scanning.

8 Data has been a part of coaching for many years. This was also the case before the data revolution, but back then the data application took place via spreadsheets and other less interactive technologies.

9 Although, there may be truth in this, the coaching and learning methodology interacts with the habits, behaviours and culture (Cortsen, Hird & Kvistgaard, 2020) and in such play a central role in influencing young players to reflect and learn.

10 A coach's role as a leader and motivator is vital – also when it comes to inclusion of young players and their individual development and supporting this mentally.

11 Cf. later section on deliberate play.

12 A player's learning in football is also a matter of obtaining a realistic understanding of his/her qualities and what he/she can work on to improve.

13 Deliberate play lets players explore though playful and enjoyable and less structured activities than deliberate practice and there is also learning to be found in unfolding talent via deliberate play.

14 This will boost the development of technical skills.

15 There may be challenges in relation to technical problems as well as the accessibility to the system, e.g., the lights, the iPad etc. One of the challenges is also that the system is very dependent on coaches and therefore clubs should consider this issue and find applicable solutions, cf. quote and discussion in a later section.

16 This refers to the feasibility and thus thoughts regarding the optimal application of the training system.

17 The fact that setting up Goal Station might be a hassle is something that clubs must consider so that they go through a proper strategic onboarding process to make sure that the investment in the system pays off in an optimal way in terms of learning how to use it, meeting the upfront costs while minimizing opportunity costs and maximizing the learning curve for the greatest number of people.

18 This relates to the fact that the surplus is often a function of mastering diverse skills, e.g., technical, tactical, physical, social, and mental skills, and that it may be 'easier' to be in possession in football.

19 When the Danish football star Michael Laudrup chipped the ball to teammate and striker Ebbe Sand during the 1998 World Cup in France in a knock-out game against Nigeria, it was a moment to remember for football lovers. He did it while running full speed and looking in another direction, and this is an example of freeing come capacity, but also depending on sublime technical skills and automatization (cf. you have done it before, cf. a similar pass to Romario for FC Barcelona).

20 Central midfielder at the professional men's team. His name has been changed to Player X to anonymize him.

21 Development of a football-specific communication ability helps to augment team-related tactical game-intelligence and it is fruitful to train this ability as 'practice makes perfect'.

22 This is interesting in relation to the former footnote about deliberate play versus deliberate practice and how these concepts influence learning for players.

23 In a recent book, researchers (Rasmussen, Rossing, Cortsen & Byrge, 2021) present an example where some of the best handball players in Denmark, e.g., Anja Andersen, who became the best in the world, benefitted from a deliberate play approach as their parents managed sports facilities where they could enjoy 'free play' with the ball.

24 In a critical reflection, it is interesting to assess if this can only be facilitated by the Goal Station system as the respondent works for the company?

25 Keep Ericsson's phrase regarding 10,000 hours in mind and combine this with 'practice makes perfect' and there are notions of importance in terms of maximizing learning outcome and effectiveness in training sessions.

26 It is a matter of making use of the club's resources.

27 In football, technical skills and tactical knowledge improves football-specific competencies as seen in the former example with Michael Laudrup and Ebbe Sand.

28 Speed must not occur at the cost of quality and lead to players running out of the stadium'. There are different technical skills in play in sports, but it is also a matter of pragmatism and finding a way to get the job done.

29 Depending on when this occurs in the lifetime of a football player. Deliberate play is very important in the younger years.

30 Via deliberate play, learning takes place through playful, creative and enjoyable activities. This is where gamification proves relevant in football training and in this Goal Station case.

31 Cultural changes have taken place over time. Growing up in the 1980s, there was not the same dependency on computers and technology, and it became a matter of figuring out how interactive technologies and data can support learning and football development.

32 This refers to the individual unfolding of deliberate play and talent.

33 This offers a polarized perspective to deliberate play, e.g., deliberate practice, and the respondent highlights how kids decades ago (before the IT and technology wave) spent more time training on their own although this statement may seem a bit vague in that is overemphasizes a quantitative and not necessarily account for a qualitative approach to training.

34 Here, it is important to include the reliability perspective in the sense that there are up to 16 repetitions in many drills and when you follow a player over a long period of time, it will reveal certain patterns that are relevant. However, coaches are of course important sparring partners for the players, and it is therefore vital to find the balance between training and contextual football relevance, cf. the quality discussion from Cortsen & Rascher (2018).

35 Conformity often characterizes traditional football training as there is a tendency that many players or coaches are resistant to too much diversity because they think that this may disturb the performance level if there are big gaps between the quality levels of players. However, this training methodology allows players with different quality levels to enjoy the same training to a higher extent.

36 I.e., deliberate practice.

37 In Kenneth Cortsen's UEFA A-license thesis, two top-notch professional players, e.g., played for clubs such as FC Barcelona, Real Madrid, Tottenham, and Ajax Amsterdam, whom he interviewed, remarked that split-vision is very important in terms of handling the ball and how to position yourself as a player.

38 When talking about players, you often see footballers, who don't have the basic skills, and then they cannot run with the ball by their feet and simultaneously keep an eye on the surroundings so Goal Station can assist in building the foundation for these skills.

39 Goal Station provides player data such as precision (whether you hit the rebounders or not), time and speed (how fast can you go through each drill while performing) so that players via the App can track their training sessions over time, check their improvement and best results, compete with friends and even watch their global ranking. These data are transferred to the App immediately after each drill.

References

Arias, P. P. (2019). *Strategic Management of Training Facilities for Success in Elite Sport Case Study of* (Doctoral dissertation, 서울대학교 대학원).

Baker, J., Côté, J., & Abernethy, B. (2003). Sport-specific practice and the development of expert decision-making in team ball sports. *Journal of Applied Sport Psychology*, *15*(1), 12–25.

Benabou, R., & Tirole, J. (2003). Intrinsic and extrinsic motivation. *The Review of Economic Studies*, *70*(3), 489–520.

Berry, J., Abernethy, B., & Côté, J. (2008). The contribution of structured activity and deliberate play to the development of expert perceptual and decision-making skill. *Journal of Sport and Exercise Psychology*, *30*(6), 685–708.

Blumer, H. (1986). *Symbolic interactionism: Perspective and method*. Berkeley, CA: University of California Press.

Brown, M. (2008). Comfort zone: Model or metaphor? *Journal of Outdoor and Environmental Education*, *12*(1), 3–12.

Busarello, R. J. (2016). *Gamification: Principles and strategies*. Sao Paulo: Pimenta Cultural.

Cairrao, M. R. (2020). *Gamification as a tool for teaching and learning futsal – Technology as a support to sports education*. Mauritius: Sciencia Scripts.

Chyung, S. Y., Stepich, D., & Cox, D. (2006). Building a competency-based curriculum architecture to educate 21st-century business practitioners. *Journal of Education for Business*, *81*(6), 307–314.

Cope, E., Partington, M., Cushion, C. J., & Harvey, S. (2016). An investigation of professional top-level youth football coaches' questioning practice. *Qualitative Research in Sport, Exercise and Health*, *8*(4), 380–393.

Cortsen, K. (2010). *10'erens Boldmodtagelse – en afgørende faktor for iscenesættelse af det offensive spil*. DBU A-træner opgave.

Cortsen, K. H., Hird, J., & Kvistgaard, P. (2020). Målet om godt købmandskab i fodboldens kompleksitet og sammenhængskraft: AaB som case for den moderne fodboldklub. In *Godt Købmandskab i det 21. århundrede* (pp. 209–254). Aalborg Universitetsforlag.

Cortsen, K., & Rascher, D. A. (2018). The application of sports technology and sports data for commercial purposes. In *The use of technology in sport - Emerging challenges*. London, England: Intech Open.

Côté, J. (1999). The influence of the family in the development of talent in sport. *The Sport Psychologist*, *13*, 395–417.

Côtè, J., Baker, J., & Abernethy, B. (2003). From play to practice: A developmental framework for the acquisition of expertise in team sports. In J. L. Starkes & K. A. Ericsson (Eds.), *Expert performance in sports: Advances in research on sport expertise* (pp. 89–115). Champaign, IL: Human Kinetics.

Côté, J., Baker, J., & Abernethy, B. (2007). Practice and play in the development of sport expertise. *Handbook of Sport Psychology*, *3*, 184–202.

Côté, J., & Hay, J. (2002). Children's involvement in sport: A developmental perspective. In J. M. Silva & D. Stevens (Eds.), *Psychological foundations of sport* (2nd ed., pp. 484–502). Boston, MA: Merrill.

De Bosscher, V. (2015). Theory of sports policy factors leading to international sporting success (SPLISS). In George B., Cunningham, Janet S., Fink , & Alison, Doherty (Eds.), *Routledge handbook of theory in sport management*.

Deci, E., & Ryan, R. (2007). Active human nature: Self-determination theory and the promotion and maintenance of sport, exercise and health. In M. S. Hagger, &

N. L. D. Chatzisarantis (Eds.), *Intrinsic motivation and self-determination in exercise and sports* (pp. 1–19). Leeds, UK: Human Kinetics.

Dewey, J. (1916). The pragmatism of Peirce. *The Journal of Philosophy, Psychology and Scientific Methods, 13*(26), 709–715.

Ericsson, K. A. (1998). The scientific study of expert levels of performance: General implications for optimal learning and creativity. *High Ability Studies, 9*(1), 75–100.

Ericsson, K. A., Krampe, R. T., & Tesch-Römer, C. (1993). The role of deliberate practice in the acquisition of expert performance. *Psychological Review, 100*(3), 363.

Fahrner, M., & Schüttoff, U. (2020). Analysing the context-specific relevance of competencies–sport management alumni perspectives. European Sport Management Quarterly, 20(3), 344–363.

Festinger, L. (1957). *A theory of cognitive dissonance.* Stanford, CA: Stanford University Press.

Giulianotti, R. (1999). *Football: A sociology of the global game.* Oxford: Black-well Publishing Ltd.

Goal Station. (2021). *Goal Station – The ultimate training* system. Retrieved from https://goal-station.com/.

Gratton, C., & Jones, I. (2014). *Research methods for sports studies.* London: Routledge.

Gréhaigne, J. F., Griffin, L. L., & Richard, J. F. (2005). *Teaching and learning team sports and games.* New York, NY, USA: Psychology Press.

Hayes, J., Rose-Quirie, A., & Allinson, C. W. (2000). Senior managers' perceptions of the competencies they require for effective performance: Implications for training and development. *Personnel Review, 29*(1), 92–105.

Hung, D. & Chen, D. (2001). Situated cognition, Vygotskian thought and learning from the communities of practice perspective: Implications for the design of web-based e-learning. *Education Media International, 38*(1), 4–11.

Hung, D. & Chen, D. (2002). Two kinds of scaffolding: The dialectical process within the authenticity-generalizability (A-G) continuum. *Education Technology & Society, 5*(4), 148–153.

Kapp, K. M. (2012). *The gamification of learning and instruction: Game-based methods and strategies for training and education.* San Francisco, CA, USA: John Wiley & Sons.

Kuper, S., & Szymanski, S. (2012). *Soccernomics: Why England loses, why Germany and Brazil win, and why the US, Japan, Australia, Turkey–and even Iraq–are destined to become the kings of the world s most popular sport.* New York, NY, USA: Nation Books.

Kvale, S., & Brinkmann, S. (2009). *Interviews: Learning the craft of qualitative research interviewing.* Los Angeles, CA, USA: Sage.

Lave, J., & Wenger, E. (1991). *Situated learning: Legitimate peripheral participation.* Cambridge, England: Cambridge University Press.

Lunce, L. M. (2006). Simulations: Bringing the benefits of situated learning to the traditional classroom. *Journal of Applied Educational Technology, 3*(1), 37–45.

McDermott, J. J. (1981). *The philosophy of John Dewey* (Vol. 1 & 2). Chicago, IL: University of Chicago Press.

McGuckian, T. B., Cole, M. H., Jordet, G., Chalkley, D., & Pepping, G. J. (2018). Don't turn blind! The relationship between exploration before ball possession and on-ball performance in association football. *Frontiers in Psychology, 9,* 2520.

Mead, G. H. (1934). *Mind, self and society: From the standpoint of a social behaviorist*. Chicago, IL: University of Chicago Press.

Piaget, J. (1977). *The development of thought* (A. Rosin, Trans.). New York: Viking Press.

Pink, D. H. (2009). *Drive: The surprising truth about what motivates us*. New York: Penguin Group.

Rasmussen, L. J. T., Rossing, N. N., Cortsen, K., & Byrge, C. (2021). Kreativitetstræning I fodbold – nye værktøjer *til spillerudviklingen*. Work in progress.

Ryan, R. M., & Deci, E. L. (2000). Intrinsic and extrinsic motivations: Classic definitions and new directions. *Contemporary Educational Psychology, 25*, 54–67.

Senge, P. (1990). *The fifth discipline: The art and practice of organizational learning*. New York: Doubleday.

Spearman, W. (2018, February). Beyond expected goals. In *Proceedings of the 12th MIT SLOAN Sports Analytics Conference* (pp. 1–17). MIT, Boston.

Trowbridge, M. H., & Cason, H. (1932). An experimental study of Thorndike's theory of learning. *Journal of General Psychology, 7*, 245–288.

Werner, K., & Dickson, G. (2018). Coworker knowledge sharing and peer learning among elite footballers: Insights from German Bundesliga players. *Sport Management Review, 21*(5), 596–611.

10 Technology, Disability and High-Performance Sport: A Socio-Cultural Reading

P. David Howe and Carla Filomena Silva

Over the last half century there has been an increasing interest in provision for sports programmes designed for populations who experience disabilities. Adapted physical activity (APA), as the academic and practical field is widely known, includes not only the high-performance end of the sporting spectrum, most notably the Paralympic Games, but also grassroots activities in schools and clubs where participation in physical activity is the focus.

In writing a chapter concerning movement technologies and disability, we have decided to focus on high-performance parasport[1], as this realm enjoys the greatest public awareness regarding disability and movement cultures. Readers should remember however that, to a greater or lesser extent, the issues highlighted within this chapter have an impact right across the APA spectrum. To start this exploration, it is important to familiarise ourselves with APA historical antecedents and the role of technology in its development.

10.1 The Role of Technology in the Development of Adapted Physical Activity

When we think of movement technologies as they relate to people who experience disability, we often think of hospital style wheelchairs that are the ubiquitous symbol of disability as portrayed on accessible parking spaces and public convenience. We also might be reminded of the wily old sea dog – Captain Long John Silver, of Treasure Island fame who used a wooden leg to walk. Wheelchairs did not appear as a movement technology until the mid-17th century, for people who were sick or experienced a movement disability (Woods & Watson, 2004). Following the First World War, Canadian physician and physical educator Robert Tait McKenzie wrote *Reclaiming the Maimed* (1918), a foundational text for the disciplines of physiotherapy. Here, he stresses how imperative it is to get the injured (or *maimed*, to use his term) to move. To achieve this goal, various therapeutic techniques needed to be adopted, many of which required the use of movement technologies.

In this chapter, when we refer to movement technologies, we consider the various apparatus that allow bodies that experience disability, to move more freely, more independently from others. Of course, historically and culturally, the freedom of movement in a wheelchair for example depends on the

DOI: 10.4324/9781003205111-10

accessibility of the physical environment, requiring adjustments such as ramps and elevators to access buildings, as well as significant adjustments such a 'curb cuttings' on sidewalks that allow for movement within an urban environment. Prosthetic lower limbs users require fewer environment adjustments to move in the public space, but it has only been in the last two decades that these technologies have made significant advancements in terms of protection of the biological body, at the point where the prosthesis is attached (Howe, 2006). Until recently many prosthesis users would often revert to using a wheelchair while recuperating from pressure sores.

Traditionally, because of the adjustments required for individuals who experience disability and use movement technologies in their daily lives, they banded together to form advocacy groups as their specific requirements for participants as fully fledged citizens have been systematically ignored, pushing them to the margins of society (Oliver, 1996; Thomas, 2007). The long term and systemic oppression of people who experience disability has led to the increased radicalisation of the disability political movement, spreading the message that individuals with impaired bodies are *dis-abled* by society, not by their impairments (see Oliver, 1990). Since the Second World War, many individuals experiencing disability have used the practice of physical activity and sport as a catalyst for rehabilitation and participation in broader social contexts (Guttmann, 1976; Howe, 2008). While the high performance practice of parasport has garnished more attention over the last 20 years, the bodies of athletes who experience disability have continually been judged in relation to an able-bodied[2] 'norm' and the standards of play and performance are compared with those of mainstream competitions. As DePauw suggests,

> It is through the study of the body in the context of, and in relation to, sport that we can understand sport as one of the sites for the reproduction of social inequality in its promotion of the traditional view of athletic performance, masculinity, and physicality, including gendered images of the ideal physique and body beautiful.
>
> (1997, p. 420)

Therefore, most users of movement technologies are unable to live up to the able bodied ideal. Of course, there are many so called able-bodied individuals who also fail to meet the lofty goals but their difference from the ideal is less evident, in part because they do not stand out as users of movement technologies. Shogan suggests:

> When persons with disabilities use technologies to adjust the participation in "normal" physical activity, the use of these technologies constructs this person as unnatural in contrast to a natural, nondisabled participant, even though both nondisabled participants and those with disabilities utilize technologies to participate.
>
> (1998, p. 272)

Individuals who experience disability must overcome the tension between the use of technology that makes them stand out individuals and the desire to participate in sport and physical activity (Seymour, 1998). Because sport is an embodied practice in which ideals of 'perfection' are celebrated, people who possess 'less than normal' bodies may shy away from the masculine physicality associated with sport. To facilitate a fair competition within high performance parasport, classification practices and regulations are used to manage, label and categorise bodies with different levels of ability (Howe & Jones, 2006; Howe, 2008). Yet, it is not only the complex classification system that separates APA activities from mainstream sport cultures, but also the adoption of performance – specific movement technologies that are used by some classified groups.

Technologies such as racing wheelchairs and prostheses have enhanced the performances of athletes for whom the loss of function deriving from specific impairments can be restored by the use of a replacement technology. The field of high-performance parasport has clearly benefited from an increase in technologies developed to harness the power of the human body (Burkett, 2010). Able-bodied high-performance athletes rely on technology in their day-to-day training (Hoberman, 1992; Shogan, 1999), yet when these athletes perform in sports like track and field athletics, the technology that has allowed them to train and compete in the sporting arena is obscured from the public eye. Able-bodied athletes take technology with them all the way to the start of an Olympic final, as their clothing and footwear are highly technological products. Butryn (2002, 2003) has highlighted that high performance (able-bodied) track and field athletics is surrounded by technology that enables athletes to become cyborgs. However, in comparison to racing wheelchairs and prosthetic limbs, specialist clothing and shoes appear less advanced, as they do not explicitly, visibly and 'unnaturally' replace parts of the biological body to aid and often enhance performance. We would argue that, just as the technologies used by able-bodied athletes seem to disappear as they blend in with their bodies in the performing act, the less subtle movement technologies used by para-athletes also become part of their embodied selves, to which we now turn our attention.

10.2 Re-embodiment

Far from being consensual, the concept of embodiment is differently defined by distinctive fields of study. In this chapter, we consider embodiment as a multidimensional and complex process in which one's bodily existence (as a fleshy, cognitive, socio-cultural and political reality) shapes and is shaped by its social, cultural and environmental context, through experience. To explore the concept of re-embodiment, we draw upon Merleau-Ponty's conceptualisation of embodiment: "a grouping of lived-through meanings which moves towards its equilibrium" (1945/1962, p. 288), foregrounding the importance of lived experience in the development of this equilibrium between the different elements of the 'grouping'. Merleau-Ponty rejected the dichotomization and fragmentation of Western scientific thought, blurring the boundaries between traditional zones of mutual exclusivity: nature/nurture, body/soul/, subject/object, etc (Iwakuma, 2002). Following Csordas, for whom embodiment is "an indeterminate methodological field defined by perceptual experience and mode of presence and engagement in the world" (1994, p. 12), inanimate and handmade objects can become incorporated into a person's body image and motor schema, such as the cane for a visually impaired person moving through spaces (Iwakuma, 2002). The incorporation of technologies into one's body can go as far as becoming integral to someone's identity: "people's peculiarities, obsessions and mixed feelings towards their aids [which] cannot be explained satisfactorily if they are seen as mere instruments" (Iwakuma, 2002, p. 79). This can clearly be seen in the attachment that some Paralympic athletes have to their mobility aids/sporting equipment (Howe, 2011). But because technologies as well as biological bodies are also autonomous objects, existing independently of this relationship, McLeod and Hawzen describe embodiment as the "paradoxical state of being at once separate and conjoined" (2020, p. 97). Such understanding is useful to make sense of the experience of *being one* with technology, for instance, when wheelchair racers perceive the wheelchairs as extensions of their agential bodies but also when, in other contexts, the autonomous existence of the chair becomes evident as a tool external to one's existence.

While different philosophical and disciplinary traditions conceptualise embodiment differently, the process always implies a relationship with something at once external to one's body, yet a relationship that becomes impressed in some significant way in one's own body. This process does not just "happen", as "human beings are actively involved in the development of their bodies over their own life cycle" (Seymour, 1998, p. 10). Within the field of mainstream sport, there is an expectation that the process of embodiment involving interactive technologies expresses itself in the form of performance improvement. For athletes who experience disability, this process of embodiment can happen very early on, such as when prosthetic legs or wheelchairs are used from a very early age, for instance in presence of congenital impairments. Yet, for most disability sport athletes, the relationship with technological objects is forced upon them, following an accident or acquired injury that significantly changes the way they are *in-the-body-in-the-world*. Because this process happens after they had already developed habitual ways of being and engaging with the world, we use the concept of *re-embodiment* to signify the ways in which "damaged" bodies re-constitutive themselves, in the context of "crisis, danger, fear, uncertainty and risk" (Seymour, 1998, p. xv). Re-embodiment is an extremely significant process, which usually involves the reconstitution of self-identity as a 'disabled' person (Seymour, 1998). The individuals' relationship with technological aids is fraught with contradictory meanings and emotions; on one hand, available technologies offer the hope of re-approximation to a "lost" and familiar state of being; on the other hand, they stand as constant reminders of *who they once were* and *what they cannot do any more* (Silva, 2013). Culturally, disability technologies such as wheelchairs and prosthetic limbs assign new social cultural meanings to one's existence, as they make one's new condition very visible to oneself and others. The use of these technologies, more often that not (it is to some degree possible to 'hide' prosthetic devices), deny the choice of 'passing' as normal in everyday interactions. Coming to terms with one's re-embodiment can be problematic. The lack of acceptance of disability was nicely articulated by Murphy (1990). In an anthropological account of slowly become paralyzed from a tumour growing on his spine, Murphy reveals how he could feel his social standing slip as he became disabled:

> A serious disability inundates all other claims to social standing, relegating to secondary status all the attainments of life, all other social roles, even sexuality. It is not a role; it is an identity, a dominant characteristic to which all social roles must be adjusted.
>
> (Murphy, 1990, p. 106)

Another important distinctive aspect of re-embodiment involving interactive technologies for people who experience disability is their compensatory, rather than augmentative qualities. While it is possible that the use of technological aids augment performance, for instance, it has been argued that specific prosthetic legs may constitute an advantage over athletes who do not wear prosthetics (Howe, 2008), the goal of technology in the process of re-embodiment is to replace and/or compensate the lost functions. Iwakuma suggests that this "embodiment cannot be complete as long as s/he is conscious of, for example, pushing a wheelchair for transportation or is making an effort to flip a page while using prosthetic arm" (2002, p. 81). Time and training are required for bodily replacement technologies to become hybridised, that is, to become so integral to one's body that it can be said to be both, biological and man-made (Howe & Silva, 2017).

The potential for hybridisation of human bodies has been encapsulated in cyborg theory. For Haraway, cyborg theory is "about transgressed boundaries, potent fusions, and dangerous possibilities" (1991, p. 154) in that "a cyborg world might be about lived social and bodily realities in which people are not afraid of permanently partial identities and contradictory standpoints" (1991, p. 154). She sees this process as a means to overcome sociocultural divisions between people.

As we will see later in the chapter, while access to expensive technology may offer Paralympic athletes more opportunities to reconfigure one's identity in positive ways, the process of re-embodiment is nevertheless challenging. Parasport athletes are not immune to a deeply ableist world, and so they can be celebrated within disability sport circles, but still shunned outside of it. This process may be even more problematic for women, as ideals of masculinity and femininity conspire against women in rehabilitation settings (Seymour, 1998). Because men are more likely to suffer spinal injuries, they outnumber women in rehabilitation centres thus the culture of rehabilitation is "dominated by masculine ideas and values ... and rehabilitation projections reflect fixed and static views of men's and women's roles" (Seymour, 1998, p. 113). This is culturally significant to parasport, as women with impairments are less likely to engage in the practice of sport because of these barriers (DePauw, 1997). While there is no space in this chapter to explore the intersectionality of Paralympic bodies, we must be mindful that these hybridised bodies have more to their identity than the experience of disability.

In this section, we explained how the process of embodiment, central to one's engagement with interactive technologies acquires distinctive contours within a context of acquired disability. Thus, the significance of this technology to the individuals themselves, their identities, everyday lives and sporting performances, is also very distinctive from the one of mainstream athletes. Next, we show how the contradictory impact of interactive technologies as simultaneously enablers of participation and enforcers of ableist divisive lines plays out within parasport contexts.

10.3 Technocratic Ideology: A Double-Edged Sword

Since the late 1980s, there has been a considered effort on the part of the IPC to force the issue of parasport into mainstream consciousness (Steadward, 1996). The problem is that only certain bodies who experience disability are eligible for parasport. Furthermore, amongst those eligible there is strong hierarchy of acceptance, with the athletes who are highly proficient in their use of mobility technologies placed at the top (Sherrill & Williams, 1996; Howe, 2008; Silva & Howe, 2018). In other words, some individuals who experience disability are marginalised even within the confines of parasport. As such, there is a need to re-evaluate what is an acceptable sporting body. In an environment where the capabilities of the physical bodies are essential, such as sport, imperfection becomes evident. DePauw (1997) examines how sport marginalises the impaired and argues that we need to re-examine the relationship between sport and the body as it relates to disability.

> Ability is at the centre of sport and physical activity. Ability, as currently socially constructed, means "able" and implies a finely tuned "able" body.'... 'To be able to "see" individuals with disabilities as athletes (regardless of the impairment) requires us to redefine athleticism and our view of the body, especially the sporting body.
>
> (DePauw, 1997, p. 423)

Hahn (1984) suggests that, because parasport is based on physical tasks, those in this community who are not as physically able as others, become further marginalised. The problem is that sporting events including the Paralympic Games still glorifies certain normalised (ableist) expressions of physical ability. As a result, more severely impaired competitors are further marginalised because their bodies are the furthest away from the 'ideal' able body. These elite participants are being given sporting opportunities in sports like boccia, a game of skill similar in many respects to lawn bowling played by severely impaired athletes with cerebral palsy who use wheelchairs. The game of boccia removes the athletes from the environs of the athletics stadium and swimming pool that are the focus of most media attention during the games. This makes the more severely impaired participant liminal to the Paralympic Games (Howe, 2008). The involvement in competitive sport, of impaired performers, is in effect accepting the social definitions of the importance of physical prowess – in essence disabling them (Hahn, 1984).

Movement technologies used in parasport like wheelchairs and prosthetic limbs have to be purchased, and therefore, the parasport movement represents a developing market for the sale of technologically advanced mobility apparatus. Many of the most up-to-date replacement technologies central to this chapter are inaccessible to athletes from low-income households in the west and many low-income countries, where costs are prohibitive. In this context, parasport may be seen as technologically advanced on the one hand, but isolationist and exclusionary on the other. This, of course, is not unique to the culture of parasport, but it is something that we need to be mindful of. State-of-the-art technologies are expensive and in the world of parasport (reflecting the world outside it), there will be *haves* and the *have-nots*.

Today, many elite para-athletes work with leading wheelchair and prosthesis suppliers to ensure that their future success is based on the technologies they use, as much as on their training regimes. While established stars are often lucky enough to garnish sponsorship deals from hi-tech manufacturers, up and coming athletes who depend on mobility technologies to compete often face financial hardship to even get to the starting line. The top cyborg athletes also receive commercial reward for their involvement in the development and manufacture of state-of-the-art technology at the heart of technocratic ideology (Howe, 2008). In other words, technology is literally as well as figuratively pushing parasport forward. As Charles suggests:

> Technology and kinesiology are symbiotically linked. They have a mutually beneficial relationship. As technology advances, so does the quality of scientific research and information accessible in the field. As kinesiology progresses and gains academic acceptance and credibility, technology assumes a more central role in our field. The more scientific the subdiscipline, the more we can see technology at play.
>
> (1998, p. 379)

Technologies such as racing wheelchairs and prostheses have enhanced the performances of athletes for whom the loss of function deriving from specific impairments can be restored by technology. The perception of success in their hybridisation depends upon the extent to which their performances are perceived as 'normal'. This normalisation is always underpinned by an 'overcoming' of disability.

> A winning wheelchair athlete is seen as the epitome of rehabilitative success. The vision of strong male bodies competing for honours on the sports field is an image that has currency in the able-bodied world. Bravery in overcoming the catastrophe of a damaged body is a quality everyone can admire.
>
> (Seymour 1998, p. 119)

The same image can be extended to amputee athletes who either through birth 'defect' or through suffering a traumatic injury require the use of performance-enhancing prosthetic limbs. The use of these mobility replacement technologies, as explained earlier provides an opportunity for re-embodiment, for the emergence of a new hybrid body (Seymour, 1998), a process unavailable to most congenitally impaired individuals. That is, users of both wheelchairs and prosthetic limbs who have acquired their impairment have the opportunity to establish a distinctive identity with their new *cyborg* bodies. These bodies are central to the public understanding of parasport.

10.4 Celebrated Parasport Hybrid Bodies

The bodies that are celebrated within the Paralympic movement – highly functioning wheelchair racers (Howe and Parker, 2012) or those who use technologically advanced prosthetics (Howe, 2011) – have increasingly become high profile because the technology they use enhances their 'normality'. Those bodies that do not use movement technologies to compete in parasport still benefit from advances in sport science support, such as biomechanical and physiological analysis, but are often still marginalised. For visually impaired, ambulant cerebral palsy and intellectually impaired athletes, who are able to compete in sport without the use of special technologies of mobility, their apparent 'normality' seems to be detrimental to how they are perceived and treated both inside and outside parasport.

Athletes with vision impairment are relatively easily understood by the public, given that a high percentage of the world's population use either spectacles or contact lenses. As our eyesight deteriorates as a result of spending too much time at the computer or through the passage of time and ageing, we can understand and appreciate the difficulties associated with poor sight. As a result, athletes with vision impairment are not treated as marginal to the same extent as those who have cerebral palsy or an intellectual impairment (Sherrill & Williams, 1996). Neurological disabilities are usually more difficult to understand than others. The uncontrollable spasticity of an individual with cerebral palsy or the intangibility of in-tellectual impairments makes it difficult for these athletes to be celebrated by the media in the way in which cyborgs are. Mobility technological inter-vention has a minimal role to play in managing these types of bodies to approximate themselves to a norm that is thought of as acceptable to mainstream sporting cultures. As a result, it is rather difficult for the general public to perceive performance excellence in some of the performances of individuals with these types of impairment.

Following Shogan (1998), it could be argued that the mobility technology used in parasport is unnatural because it makes athletes less than human. In fact, in the lead up to the London 2012 Paralympic Games, a television sta-tion in the United Kingdom ran an awareness campaign entitled 'Freaks of Nature' (Silva & Howe, 2012). This campaign was designed to highlight the 'supercrip' in parasport, but it did not translate well. For those whose bodies are explicitly cyborgs, the 'super-human' results achieved through the use of either state-of-the-art wheelchairs or prosthetic limbs within Paralympic track and field athletics have become the new norm and accepted currency over the last two decades within the public understanding of ability within parasport. Replacement technology allows for exceptional sporting performances cele-brated by the able-bodied public, but such performances are unlikely to be achieved by athletes who compete without these mobility aids. This use of what Butryn (2003) coined 'implement technology' (performance-enhancing) has transformed parasport into a significant sporting spectacle.

10.5 Discussion

In an increasingly commercial and interconnected world, the technocratic ideology (Charles, 1998) apparent at the Paralympics Games and parasport more generally will be hard to challenge. The athletes who use wheelchairs and prosthesis are at the centre of the Paralympic Games and will be better advertisers for consumers simply because their sporting excellence can high-light specialist materials for others to purchase if they wish to compete at the highest level. The body policing on what is acceptably human and what is not (Cole 1993, 1998), evident in mainstream high-performance sport, has been reversed in the world of parasport. In Paralympic track and field athletics, the closer a body is to a cyborg the more capital it holds, leading to a hierarchy of disability acceptability that places those bodies whose disability cannot be compensated by technological apparatus as less acceptable. This defies Haraway's (1991) utopia that the blurring of boundaries between humans and non-humans could create a more inclusive world.

Parasport wheelchair users and amputees who use prostheses have successfully re-embodied, and therefore blurred the lines between 'natural' and 'artificial'. They are perhaps the best example of the cyborg in contemporary society.

These athletes exhibit what Butryn (2003) coined *hegemonic humanness* that is the nexus between the natural and legal and the artificial and illegal. Hegemonic humanism can be seen to have been practiced when Oscar Pistorius was initially excluded from competing in able-bodied athletics (Howe, 2008, 2011). The restoration of his right to compete on his prosthesis was restored because he had no other option but to run on man-made legs and by the fact it was concluded by scientists that they were not advantaging him in the context of competition. In a sense, parasport celebrates "transgressing the taboo boundary between blood, sweat, and tears, and blood, sweat and gears" (Butryn, 2003, p. 28). The cyborg wheelchair user and the prosthetic limb wearer are the parasport movement triumphs. This is largely because the Paralympics Games were designed to celebrate a corporeal entity that is distinct from the able-bodied norm. Yet, it appears that Paralympic distinctiveness must increasingly take on a cyborg form, rather than showcase human bodily diversity.

Where does this leave 'les autres', the athletes whose bodies do not fit neatly into the categories that can be enhanced by mobility technology? They certainly have a part to play in the Paralympic movement, but the more marginal the physicality of the body, the further away it is from the potential of cyborgification (Howe & Silva, 2017) and the more likely that a tragic rather a heroic allegory will follow them. This analysis tells us a great deal about the politics of disablement. While it is considered an infringement for the able to become too cyborg, for people who experience disability it is highly advantageous because technology can normalise their 'inferior' bodies to the point where they can produce super-human results. Of course, there is a tension here. MacIntyre (1999) tells us that vulnerability and affliction and the related quality of dependence are central to the human condition. The susceptibility to injury and misery, distress and pain is likely to befall on us at some point in our existence. We all will be reliant on others from time to time. It begs the question of why bodies that experience disability are so harshly treated by society and, at least in the context of parasport, only those who are cyborgs are celebrated at length? Of these cyborgs, winners are held up on a pedestal as supercrips (Howe, 2008; Silva & Howe, 2012) as testament to the hopeful potential of a symbiotic relationship with technology to conquer the vulnerable and frail qualities of human existence.

10.6 Conclusion

In the last 30 years, the associated development of biomechanically and ergonomically responsive prostheses has meant that many athletes who in the past would have competed from a wheelchair are now able to compete from a standing position. While the development of replacement mobility technology that enhances sport performance is understandably beneficial for those who can go through the process of cyborgification, it marginalises further those athletes who do benefit from these in their competitive performance. Because high-end wheelchair athletes are able to perform at the same level or better than able-bodied athletes, the abilities of these athletes are obvious to the public. On the other hand, it might be difficult to see the ability of an athlete whose cerebral palsy affects both legs and runs 100m much slower than his/her able-bodied counterpart.

The possibility of a re-embodiment for certain athletes who experience disability is provided through acquiring expensive sporting technologies central to the process of cyborgification, while excluding many others through lack of financial means. In parasport, there are increasing numbers of athletes with mechanical, artificially designed, hybrid bodies creating new sporting potential. The technology they use has the capacity to 'normalise' their bodies, and in so doing produces 'sporting cyborgs', who are celebrated both inside and outside the parasport movement, because they increase its marketability and the public's awareness of the ability of certain impairment groups. A technocentric ideology underpins the cyborgification process celebrated within parasport and has made celebrities of the athletes who are successful in using the state-of-the-art replacement technologies to achieve super performances (Howe 2008, 2011). Regrettably, such elevated status of handpicked cyborgs can be problematic for the communities of impaired individuals who can never achieve such a position. As Kama argues:

> (w)ell-known, successful disabled people are put on a pedestal for their demonstrated ability to triumph. This triumph is used to validate the disabled individual and to alter societal perceptions. Consequently, the wish to see disabled who 'have done it' is particularly intense while the pitiful disabled trigger antipathy because they reproduce and reinforce disabled people's inferior positionality and exclusion.
>
> (2004, p. 447)

Ultimately, the Paralympic Games may become a replacement technology showcase, rather than a contest of athleticism, leaving behind those who cannot either afford or use performance-enhancing technology. In short, technological advancement in relation to parasport is not dissimilar to other changes in society; it is clearly a double edge sword:

> Prosthetic medicine is dedicated to physical normalisation and is devoted to the artificial alteration of both function and appearance, but it enters the realm of biopolitics because it uses the 'normal' body as its tribunal and blueprint for action, and treats the impaired body as a spoilt entity that must be hidden and corrected.
>
> (Hughes, 2000, p. 561)

Technology empowers some, while leaving the status of others at best unaltered, or, at worst increasing their liminality. While Haraway (1991) believes that cyborgization can bring more people into the fold of the 'humanist subject', maybe the whole notion of what constitutes a humanist subject must be rethought. This is because

> The celebration of the... cyborg bodies of the best wheelchair and prosthetic-wearing athletes, is good for the individuals placed on the pedestal but may lead to the (dis)empowerment of other athletes with impairment who cannot take advantage of the explicit use of technology. Ultimately the Paralympics risk becoming a show of technology, rather than a show of athleticism.
>
> (Howe, 2011, pp. 879–880)

Notes

1 Parasport refers to sport that is practiced under the governance rules of the International Paralympic Committee (IPC).
2 We use the term 'able-bodied' here because it is the term used by athletes within the cultural context of the Paralympic Games.

References

Burkett, B. (2010). Technology in Paralympic Sport: Performance Enhancement or Essential for Performance. *British Journal of Sports Medicine, 44*(3), 215–220.

Butryn, T. (2002). Cyborg Horizons: Sport and the Ethics of Self-Technologization. In Andy Miah & Simon Easson (eds.), *Sport, Technology: History, Philosophy, and Policy* (p. 111–134). Oxford: Elsevier Science.

Butryn, T. (2003). Posthuman Podiums: Cyborg Narratives of Elite Track and Field Athletes. *Sociology of Sport Journal, 20*(1), 17–39.

Charles, J. (1998). Technology and the Body of Knowledge. *Quest, 50*(4), 379–388.

Cole, C. (1993). Resisting the Canon: Feminist Cultural Studies, Sport, and Technologies of the Body. *Journal of Sport and Social Issues, 17*(2), 77–97.

Cole, C. (1998). Addiction, Exercise, and Cyborgs: Technologies and Deviant Bodies. In Geneviève Rail (ed.), *Sport and Postmodern Times* (pp. 261–275). Albany: State University of New York Press.

Csordas, T. (1994). *Embodiment and Experience: The Existential Ground of Culture and Self.* Cambridge: Cambridge University Press.

DePauw, K. (1997). The (In)Visibility of Disability: Cultural Contexts and "Sporting Bodies." *Quest, 49*(4), 416–430.

Guttmann, L. (1976). *Textbook of Sport for the Disabled.* Aylesbury: HM+M.

Hahn, H. (1984). Sports and the Political Movement of Disabled Persons: Examining Nondisabled Values. *Arena Review, 8*, 1–15.

Haraway, D. (1991). *Simians, Cyborgs, and Women: The Reinvention of Nature.* London: Routledge.

Hoberman, John. (1992). *Mortal Engines: The Science of Human Performance and the Dehumanization of Sport.* Oxford: The Free Press.

Howe, P. D. (2006). The Role of Injury in the Organization of Paralympic Sport. In S. Loland, Skirstad, & I. Waddington (eds.), *Pain and Injury in Sport: Social and Ethical Analysis* (pp. 211–225). London: Routledge.

Howe, P. D. (2008). *The Cultural Politics of the Paralympic Movement: Through the Anthropological Lens*. London: Routledge.

Howe, P. D. (2011). Cyborg and Supercrip: The Paralympics Technology and the (Dis) Empowerment of Disabled Athletes. *Sociology, 45*(5), 868–882.

Howe, P. D., & Jones C. (2006). Classification of Disabled Athletes: (Dis) Empowering the Paralympic Practice Community. *Sociology of Sport Journal, 23*(1), 29–46.

Howe, P. D. & Parker, A. (2012). Celebrating Imperfection: Sport, Disability and Celebrity Culture, *Celebrity Studies, 3*(3), 270–282.

Howe, P. D. & Silva, C. F. (2017). The Cyborgification of Paralympic Sport. *Movement & Sport Sciences - Science & Motricité*, (97), 17–25. 10.1051/sm/201 7014

Hughes, B. (2000). Medicine and the Aesthetic Invalidation of Disabled People. *Disability and Society. 15* (4), 555–568.

Iwakuma, M. (2002). The Body as Embodiment: An Investigation of the Body by Merleau-Ponty. In M. Corker & T. Shakespeare (eds.), *Disability/Postmodernity: Embodying Disability Theory* (pp. 76–87). London: Continuum.

Kama, A. (2004). Supercrips versus the Pitiful Handicapped: Reception of Disabling Images by Disabled Audience Members. *Communications, 29*(4), 447–466.

MacIntyre, Alasdair. (1999). *Dependent Rational Animals: Why Human Beings Need the Virtues*. Chicago: Open Court.

McKenzie, R. T. (1918). *Reclaiming the Maimed*. Philadelphia: W.B. Saunders.

McLeod, C. M., & Hawzen, M. C. (2020). Body Objects, Poltitical Physics,and Incorporation. In J. I. Newman, H. Thorpe, & D. L. Andrews (eds.), *Sport, Physical Culture, and the Moving Body: Materialisms, Technologies, Ecologies* (pp. 87–107). Oxford: Rutgers University Press.

Merleau-Ponty, M. (1962). *Phenomenology of Perception*. (Colin Smith, Trans.). London: Routledge. (Original work published 1945).

Murphy, R. (1990). *The Body Silent*. New York: W.W. Norton & Company.

Oliver, M. (1996). Understanding Disability: From Theory to Practice.

Oliver (1990). *The Politics of Disablement: A Sociological Approach*. St. Martins Press.

Seymour, W. (1998). *Remaking the Body: Rehabilitation and Change*. London: Routledge.

Sherrill, C., & Williams, T. (1996). Disability and Sport: Psychosocial Perspectives on Inclusion, Integration and Participation. *Sport Science Review, 5*(1), 42–64.

Shogan, D. (1998). The Social Construction of Disability: The Impact of Statistics and Technology. *Adapted Physical Activity Quarterly, 15*, 269–277.

Shogan, D. (1999). *The Making of High Performance Athletes*.University of Toronto Press.

Silva, Carla (2013). The Impact of Sitting Volleyball Participation on the Lives of Players with Impairments. Doctoral Thesis. Loughborough University. https:// hdl.handle.net/2134/14178

Silva, C. F., & Howe, P. D. (2012). The [In]Validity of *Supercrip* Representation of Paralympic Athletes. *Journal for Sport and Social Issues, 36*(2) 174–194.

Silva, C. F., & Howe, P. D. (2018). The Social Empowerment of Difference: The Potential Influence of Parasport. *Physical Medicine and Rehabilitation Clinics of North America, 29*(2), 397–408. 10.1016/j.pmr.2018.01.009

Steadward, R. D. (1996). Integration and sport in the Paralympic movement. *Sport Science Review, 5*(1), 26–41.

Thomas (2007). *Sociologies of Disability and Illness: Contested Ideas in Disability Studies and Medical Sociology.* Palgrave Macmillan.

Woods, B., & Watson, N. (2004). The Social and Technical History of Wheelchairs. *International Journal of Therapy and Rehabilitation, 11*(9), 407–410.

11 In Lieu of an Afterword: "Embodiment and Skill Acquisition in Sports Technologies" Course Syllabus

Veronika Tzankova and Michael Filimowicz

11.1 Introductory Course Rationale

Embodiment and Skill Acquisition in Sports Technologies explores the relationship between athletic skill acquisition and interactive technologies based on overlapping yet differentiated epistemological approaches coming from the scholarship in the domains of embodiment and HCI. Emerging from both of these domains' concern for experience, this course investigates the bilateral connection between (1) the potential of athletic movement expertise to expand on embodied design, and methodological approaches in the field of HCI through the integration of interactive technologies within the context of sports and (2) the capacity of interactive technologies to support not only the functional and measurable aspects of movement within athletic skill expertise, but also its experiential dimension by creating technologies for reflection and somatic awareness. Thus, Embodiment and Skill Acquisition in Sports Technologies aims at establishing a coherent theoretical ground for exploring and pushing the boundaries of technologies in relation to sports while eliciting an interdisciplinary inquiry leading to the mobilization of more richly articulated athletic movement knowledge within the areas of physical literacy, physical intelligence, perceptual-motor performance, and sports technologies (as illustrated in Figure 11.1). Students will acquire knowledge and skills needed to understand, research, and participate in the design of interactive technologies for athletic performance based on novel soma-based epistemological approaches to human-technology relations.

Movement is central to human existence. At each stage of life, we continuously try to develop new motor skills or to refine the ones we have already acquired in the hope of enhancing productivity and quality of experience. This is specifically the case within the domain of sports where athletes are evaluated almost exclusively on their ability to produce and replicate such skills, often in a broad range of performance conditions. Skilled athletes dedicate a significant amount of time in practicing and perfecting these skills with the objective of enhancing performance and achieving excellence. The achievement of excellence in sports is primarily measured through the functional aspects of movement, where its experiential, aesthetic, and reflective dimensions have not

DOI: 10.4324/9781003205111-11

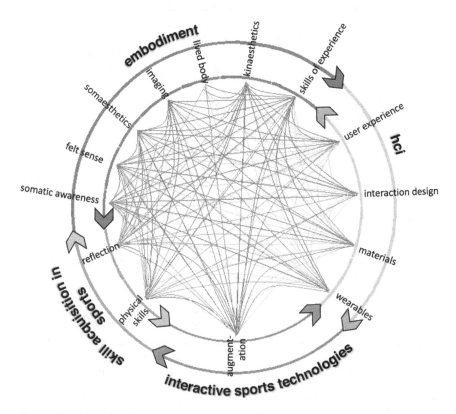

Figure 11.1 Mapping the course structure and the relations between key fields and themes.

been fully explored. This is evidenced by the scholarship on skill acquisition in sports and its strong focus on motor performance, sensori-motor skills, co-ordination, and musculoskeletal constraints among others. Regardless of the physical limits of our bodies however, movement has always been our primary vehicle of existing and sustaining a connection with the world outside ourselves (Sheets-Johnstone, 2011). It serves a highly experiential role that establishes our aesthetico-relational flow with the environment and defines our experiences.

The turn to embodiment (Dourish, 2001) within the increasingly prevalent concern for experience in the domain of HCI has initiated a growing interest in the practical application of somatic sensibilities – a body's capacity for sensori-aesthetic appreciation, including movement – to all forms of technology applications and interactions. In light of this development and accounting for the increasingly pervasive understandings of computers as "a set of invisible distributed processes" (Schiphorst, 2007, p. 4), this course explores the ways in which interactive technologies can introduce novel ways for understanding and supporting movement and skill acquisition in sports by turning toward somatics and embodiment as an "experience tradition" (ibid, p. 10). The detailed exploration of movement as a somatic and aesthetic modality in addition to its functional aspects carries a tremendous potential to enhance the flow of experience essential to both HCI and sports practice. HCI and sports are brought together in the emergent field of Sports-HCI which still preoccupies itself with the measurable parameters of movement (Mueller & Young, 2018), neglecting human factors that assist users in rediscovering themselves through reflexive experiencing of their bodies in movement.

Embodied approaches to technology design have been critiqued for a lack of integrated theories, principles, models (Moen, 2005) and contextual stability (Dourish, 2004). While the scholarship on movement-based interaction is expanding, a common ground for understanding and modeling movement as an HCI parameter has not been established. As athletic performances take place within a particular sport discipline's governing rules, these rules can serve as epistemological frameworks that inform the utilization, representation, and expression of movement in relation to interactive technologies and HCI in general. Exploring athletic skill development within its functional and experiential dimensions carries the potential to provide creative solutions to the quickly unfolding technological landscape.

This course is also about creating a reflective space and an educational opportunity to assist students – especially those who do not have a strong awareness of the tremendous experiential potential underlying movement – find points of access to it through the explorative lens of somaesthetic design and HCI. A part of the course is devoted to learning about the theories and practices regarding the body as a site of aesthetic appreciation (Shusterman, 2008), and animation as the primary form of being in the world (Sheets-Johnstone, 2011). This part is not specifically about sports or athletics; it is about learning to physically experience how our bodies mediate our understandings of the world and navigate the sensations of being alive.

Thus, this course intersects four overlapping and yet distinguished domains of practical and theoretical inquiry – embodiment, athletic skill acquisition, HCI, and interactive sports technologies. Incorporating the rich epistemologies of these, it presents a survey of the key sources and case studies which lay out the historical contexts, theoretical frameworks, and contemporary debates surrounding the notions of embodiment and skill acquisition in/through contemporary sports technologies. The course 'kicks off' with a reading from Margaret Whitehead (2010), who chronicles the historical intertwining between the ideas of embodiment and physical literacy. The reading provides a useful contextual background for the weekly key themes of the course. The course is then divided into three somewhat distinct modules representing prevalent theoretical frameworks from the perspectives of embodiment and skill acquisition, sports and HCI, and contemporary debates and considerations.

Module I introduces key themes from the interlacing domains of embodiment and skill acquisition in/through sports. This section explores the role of the body – along with its senses, perception, proprioception, kinesthesis, animation, felt sensations, flow, and lived experiences among others – in the process of skill acquisition, knowledge generation, meaning-making, and experiencing of the world. It is particularly invested in the phenomenological treatment of the body as a locus of aesthetic experiences. This module sets the context for understanding the ways in which sports and HCI have been recently co-evolving through advancements in movement knowledge expertise and creative embodied disciplines.

Module II focuses on the specific and novel integration of interactive technologies into the domain of sports. Starting with the historical trends (waves) that have shaped the intellectual trajectory of the HCI field and taking the 'turn to embodiment' as its point of departure, the module introduces insights and expertise regarding the role of movement in the formation of user experience as well as emerging soma-based design approaches and methodologies. It then extends to the practical utilization of interactive technologies in sports activities by looking at specific case studies, projects, and latest developments within the domain of Sports-HCI.

Module III targets the sharing and fostering of ideas and methods which advance research in Sports-HCI through a discussion of broader topics and contemporary debates that relate to the field. It also engages with the wider impact of such technologies on non-human agents, such as (but not limited to) animals and the environment. To exemplify the importance of ethical considerations as an integral part of the design and application of technologies, the module includes a review of the use of technologies in equitation.

11.2 Key Themes

11.2.1 Week 1 Course Overview

Introduction to the course and its learning objectives.
   Read:
Whitehead, M. (2010). *Physical literacy: Throughout the lifecourse* (1st ed).
   Routledge. – Chapter 2.

11.2.2 Module I Embodiment and Skill Acquisition in/through Sports

11.2.2.1 Week 2 Sporting the Embodied Mind

Key themes: embodiment, embodied cognition, being in the world, lived experiences in sports

We start the theoretical frame of the course with discussing the notion of embodiment as it is one of our central organizing themes. It also presents a major consideration for the theories and practice of HCI. Embodiment typically refers to our lived experiences, feelings, vitality, and visceral flows that come from having a body. This contrasts with the information processing approaches to human cognition where thinking is characterized as the milieu providing data input which then gets transformed by our senses. Theories of embodiment focus on how our bodies and experiences shape how we perceive, sense, feel, think, and understand. The notion of embodiment is explored by a number of different theoretical and practical traditions. Most notably, embodiment theorists engage with the existential and intellectual concerns surrounding the self and its fragmented, divided, non-unified cognizing subject. Recognizing the logical and experiential gaps in relation to the body-mind dichotomy prevalent in Western intellectual tradition, the notion of embodiment proposes to build a bridge between mind in science and mind in experience. It is argued that cognition should be understood as an emergent process that takes place through an organism's active engagement with the world.

Starting with a discussion of the natural history of consciousness, we will proceed to explore Varela et al.'s (1991) proposed understanding of cognition as "the enactment of a world and a mind on the basis of a history of the variety of actions that a being in the world performs" (ibid, p. 9). This understanding of cognition is then translated into the sporting body whose multi-textured lived experiences and relations to the environment are shaped – and maybe to an extent determined – by a sport discipline's governing rules. The discussion of embodiment in the specific context of sports reconsiders conceptualizations of the body as a motor-physiological tool to engage its capacity for experience, especially through pushing the boundaries of movement.

Read:

Sheets-Johnstone, M. (1998). Consciousness: A natural history. *Journal of Consciousness Studies, 5*(3), 260–294.

Varela, F. J., Thompson, E., & Rosch, E. (1991). *The embodied mind: Cognitive science and human experience.* MIT Press. – Chapters 1&2.

Allen-Collinson, J. (2009). Sporting embodiment: Sports studies and the (continuing) promise of phenomenology. *Qualitative Research in Sport and Exercise, 1*(3), 279–296.

11.2.2.2 Week 3 Somatics/Somaesthetics for Athletics

Key themes: soma, somaesthetics, aesthetics, somatic awareness and somatic reflection for athletic performance

Extending on the theme of embodiment presented in week 2, week 3 explores the notion of the soma as an essential consideration of embodiment and a point of departure for the interdisciplinary field of somaesthetics. The term *soma* refers to the body as experienced from within by first-person perception. It represents the lived body as internally sensed and defines a categorically distinct form of awareness grounded in immediate proprioception and unique sensory mode. Richard Shusterman (2008) – a pragmatist philosopher, extends on the notion of the soma to compound it with aesthetics and to develop the field of what he calls *somaesthetics*. Somaesthetics emphasizes the primacy of our bodily sensations and somatic experiences as a way of being and thinking. The experiences of the lived body are understood as form of aesthetics, where the soma is conceived as a "locus of sensory-aesthetic appreciation(aesthesis) and creative self-fashioning" (p. 19). The notion of somaesthetics expands on pragmatism's focus on agency to unmask the complex role of the body in its aesthetic experiencing of the world. Here, *aesthetic* takes the meaning of both sensing and remaking the situation, where the body is "our primordial instrument in grasping the world" (p. 19). Based on these ideas, Shusterman develops the argument that we should not only recognize the body's complex ontological structure in its material objectivity, but also in its subjective lived experiences of the world.

We will then relate the project of somaesthetics to athletic proficiency in sports. Dominant theories of motor learning have speculated that expert athletic performance depends on exerting precise control over bodily movements. This conceptual approach with practical implications has been rarely challenged. There is an emerging body of evidence, however, that somatic reflection can serve as a mediator of athletic improvement. Similar arguments (although outside the scope of this course) have been made in the field of dance, where somatic reflective practices have been established to improve a dancer's performance.

Read:

Hanna, T. (1986). What is somatics? *Somatics Journal of the Bodily Arts and Sciences*, 5(4), 4–8.

Shusterman, R. (2008). *Body consciousness: A philosophy of mindfulness and somaesthetics.* Cambridge University Press. – Introduction & Chapter 1.

Toner, J., & Moran, A. (2015). Enhancing performance proficiency at the expert level: Considering the role of 'somaesthetic awareness.' *Psychology of Sport and Exercise*, 16, 110–117.

11.2.2.3 Week 4 Movement: Kinetic-Kinaesthetic-Affective Dynamics

Key themes: animation, movement, kinaesthetics, movement that transcends the body, tactility, flow

Movement is ubiquitous in our interactions with and understandings of the world. Extending on the phenomenological treatments of the soma as a site of lived experience and aesthetic sensibilities, we are going to look at movement as a primary form of being in the world. What is livingly present in the experience of movement? What distinguishes kinaesthetic from kinetic experiences of movement? In what way is movement pertinent to receptivity and responsivity? In her work on movement and animation, Maxine Sheets-Johnstone addresses these questions. According to her, the essence of being animate is the capability of moving oneself and experience the temporal-energic dynamics of moving. To move is not enough; one has to be able to feel the dynamic flow of movement too. There is no movement without bodies, but there is – phenomenologically speaking – *movement that transcends the body*. It is this aspect of movement that we are going to explore.

Moving involves an affective impulsion of kinaesthetically experiencing the felt body in motion in a tactile way which expands to a tactile-kinaesthetic consciousness. Here *tactile* refers to Aristotle's sense of being alive and making one's way through the world as a form of phenomenological experience. This leads to the classification of movement as a *primal sensibility* – the argument that sensation "does not arise out of immanent grounds, out of psychic tendencies; it is simply there, it emerges" (Sheets-Johnstone, 2014, p. 248). Any action or activity involves movement – thus "I move" precedes "I do" and "I can." The origin of the skilled ego and the practical subjects' "I can" is always found in the experience of existential reality – the "I move." We do not come into the world knowing, but we come moving. Movement is not a matter of sensation, but of dynamics. Sensations are limited in time and space, so they lack the experience of flow. But movement is filled with flow which forms the repertoire of familiar kinesthetic flows. Flow is defined by a temporal continuity, where the only way we can feel time is through movement.

Read:

Sheets-Johnstone, M. (2011). *The primacy of movement* (Expanded 2nd ed). John Benjamins Publishing. Chapters 1&2.

Sheets-Johnstone, M. (2014). Animation: Analyses, elaborations, and implications. *Husserl Studies*, 30(3), 247–268.

11.2.2.4 Week 5 Inner Senses and Relational Flows

Key themes: feel, felt sense, flow, relational flow in sports, autotelic action

Building on the previously presented key themes, this week's content further expands on the relation between the body, perception, environment, and interaction. We will start with Eugene Gendlin's notion of the felt sense. Gendlin upholds that "something that is not as yet" performs functions – and these functions are generated by the body, not by perception. Gendlin's primary argument is that the body's interaction with the environment precedes any forms of abstract knowledge. This is illustrated by the example of situations when we feel uneasy for no specific reasons. Gendlin calls such vague, pre-verbal viscerality – *felt sense*: the inner knowledge or awareness that has not been consciously processed and cannot be assigned to a specific sense-apparatus. After absorbing all relevant information, the body has the capacity to imply a right next step beyond traditional methods of knowledge acquisition. Gendlin's conceptualization of the felt sense has been utilized in both sports and HCI (Núñez-Pacheco & Loke, 2018) as it makes pre-verbal bodily awareness more tangible. For example, the notion of feel is often used in equitation to point to the primacy of bodily experiences which cannot be put into words.

The notion of the felt sense and its pre-intentionality relates to broader forms of pre-intentional consciousness which can also reveal themselves in what the philosopher Michel Henry calls the affectivity of life – the "invisible, vital power intuited amid worldly appearances and sometimes limited bodily expressions" (via Smith & Lloyd, 2020, p. 538). Such affectivities remind us how good it is to be alive – a state called *flow*. A state of flow occurs when we experience complete absorption with the activity at hand and our actions become *autotelic* or done simply for the sake of doing, without any external motivation. Based on the notion of flow, Smith & Lloyd formulate the idea of relational flow – where we begin to appreciate the gushes, rushes, and bursts of vivacity in the interactional space of moving with an 'other'. This concept can be illustrated with examples from contact sports – horseback riding, karate, wrestling, competitive dancing, soccer, among many others – where participants necessarily come into bodily contact with one another.

Read:

Gendlin, E. T. (1993). Three assertions about the body. *The Folio, 12*(1), 21–33.

Smith, S. J., & Lloyd, R. J. (2020). Life phenomenology and relational flow. *Qualitative Inquiry, 26*(5), 538–543.

11.2.2.5 Week 6 Skill Acquisition and Embodiment

Key themes: tacit knowledge, knowledge-in-action, reflection-in action

This week engages with the idea of tacit knowledge (Polanyi, 1966) and knowledge- and reflection-in-action (Schön, 1983) as a mediary point between embodiment, felt senses, and the acquisition of practical skills. The idea of tacit knowledge was first introduced by Michael Polanyi in his work Personal Knowledge and then further developed in The Tacit Dimension. It is defined as "a way to know more than we can tell" – more precisely, the translation of bodily experiences into perception of things. By elucidating the way bodily processes participate in perception, we become capable of tracing the bodily roots of all thought. Our bodies serve as an essential tool of all external knowledge – whether it is intellectual or practical. This explains why our bodies are the only thing in the world we do not experience as an object, but rather as a relation to the external world. We 'feel' bodies as our own by intellectual contemplation. Awareness of our bodies suggests that we have a *feeling* of their existence.

Similarly, Donald Schön (1983) examines the ways in which skilled professionals' work-related activities are guided by reflective practice. Reflective practice refers to individuals' developing an awareness of their tacit understandings and ability to learn from experience – a process which Schön labels as *knowing-in-action*. By exploring the elements of knowledge-in-action, Schön demonstrates that professional knowledge is developed within action. The concept *reflection-in-action* is then invoked to denote the active and non-propositional processes in which new knowing-in-action is developed. Schön's consideration of *knowledge-in-action* and *reflection-in-action* which link problem solving skills to broader spectrums of reflective explorations and corrective measures in action are highly applicable to the study and practice of athletic performance/training as they underpin particular forms of embodied knowing realized through highly specific movements.

Read:

Polanyi, M. (1966). *The tacit dimension*. Doubleday & Company.

Schön, D. A. (1983). *The reflective practitioner: How professionals think in action*. Basic Books.

11.2.3 Module II Sports and HCI

11.2.3.1 Week 7 Embodiment and HCI

Key themes: third-wave HCI, the embodied turn, embodied knowing, embodiment through sports technologies

Building on the topics of embodiment, somatics, somaesthetics, kinaesthetics, feel, and tacit knowledge explored in Module I, this week's content presents an introductory exploration to human-technology relations through the lens of experience, embodied knowing, and coenesthesia among others. Since the 1980s, the humanities have positioned the notion of embodiment as a key frame to understanding cognition. Following this shift, HCI has similarly turned toward the lived experiences of the body emphasizing the embodied and situated nature of interaction. This new paradigm has been labeled as "third wave HCI" (Bødker, 2006). Within it, embodiment emerges as the foundation of interaction (Dourish, 2001). Embodied Interaction is a genre of interaction which aims at moving beyond the graphical user interface paradigm toward engaging the various dimensions of our lived experiences emanating from our bodies and their abilities as perceived from first-person perspective. It shifts the design focus away from systems and objects to a first-person perspective and an unfolding exploration of meaning making. The notion of embodiment in HCI has since been extended to include the soma and its aesthetic dimension of experience along with notions of embodied knowing – an active process of inquiry rooted in bodily experiences being and acting in the world. Thus, this week's topic on embodiment and HCI explores the historical context and conceptual frameworks that have made this coupling possible. It sets the ground for understanding how somatic practices have been gradually utilized as design modalities and first-person methodologies (content of week 10). While the theory and practice of embodiment is transforming HCI, we will also discuss the potential of technologies to transform embodiment itself.

Lastly, we will look into the ways in which the notion of embodiment has been utilized as a design modality and an aesthetic 'function' of interactive sports technologies through Warren et al.'s (2016) case study on the role of digital technologies in facilitating a more conscious and 'objective' evaluation of a rider's balance and position in equitation.

Read:

Bødker, S. (2006). *When second wave HCI meets third wave challenges* (pp. 1–8). ACM Press.

Loke, L., & Schiphorst, T. (2018). The somatic turn in human-computer interaction. *Interactions, 25*(5), 54–58.

Warren, J. L., Matkin, B. B., & Antle, A. N. (2016). *Present-at-body self-awareness in equestrians: Exploring embodied "feel" through tactile wearables* (pp. 603–608). ACM Press.

11.2.3.2 Week 8 Sporting Bodies in Technologies

Key themes: Sports-HCI, interactive tech and sports, bodily interaction, embodiment, materiality, information architecture, content, applications

The theme of this week explores the interception of sports and HCI. While sports and recreational fitness activities have become a rapidly expanding area within consumer-oriented digital technologies, a much richer consideration of the role of computational media and design in connection to training and safety in professional and amateur athletics beyond the consumption of popular, activity-tracking electronics has been limited. By conceptualizing how characteristics of the sports domain can further facilitate knowledge in the HCI domain and how interactive technologies can be utilized to advance athletic performance, we are going to explore the junction of sports and HCI in its potential to generate new interaction metaphors. Sports are still an underrepresented domain within the field of HCI, where existing work can be summarized under the following categories: (1) technical exploration – focusing on the technical potential of sensor-implementation (e.g., wearables) within sports domains; (2) bodily interaction – design of novel interaction based on body movement/motility; (3) new forms of play – where sports have inspired new forms of digital interactive games; and (4) socio-motivational systems – which motivate humans to move and exercise (Nylander et al., 2014). We will organize the intersection of sports and HCI with a slightly different classification than the one proposed by Nylander et al. which accounts for both users and technologies: embodiment, materiality, information architecture and design, content, and applications. Embodiment targets improved performance, body awareness and skills, as well as better relation of the body to its surrounding environment in a fast-paced real-time context. Materiality engages with the material aspects of interactive sports technologies. Information architecture and design explores the new ways of organizing and capturing sensor input, processing and output. Content deals with new ways of adding rich media content into sports equipment to support more immersive and engaging experiences in their affective and aesthetic dimensions. Applications investigate the various domains within sports where interactive technologies can be used to enhance and support performance.

Read:

Mueller, F., & Young, D. (2018). 10 lenses to design sports-HCI. *Foundations and Trends in Human–Computer Interaction, 12*(3), 172–237.

11.2.3.3 Week 9 Wearable Technologies and Sports

Key themes: wearable technologies, sports

As computation is becoming more ubiquitous and the distance between technologies and our bodies is disappearing, we are going to extend on the theme of sporting bodies in technologies by considering the ways in which wearables can support athletes while exercising. While there are numerous wearable/smart technologies in the market – such as smart watches, fitness trackers, fitness apps – these all seem to be measuring physical activities in terms of quantitative data. Such data, however, reflects only on performance, but not on technique. These also do not seem to affect the motivation for doing sports or the experience of such activities. Thus, we look at the possibilities of incorporating wearable technologies for real-time support of athletes during physical exercising. Currently, an effective analysis of techniques essential to sports performance can only be provided by professionals – usually using slow motion video analysis. Since most athletes do not have sufficient understanding of biomechanics and no access to a professional trainer, in some cases this leads to adaptation of wrong techniques. To overcome these problems, it is necessary to go beyond quantitative feedback that addresses simply the quantifiable aspects of physical movement and focus on meaningful support that emphasizes the qualitative aspects of physical activities. Systems trying to serve this purpose can benefit from considering the creative potential of movement that brings forth a reflective space for rediscovering its ameliorative and experiential properties. We are then going to consider the possibilities surrounding the location of wearables on the body from the perspective of somatic sensibilities, soma design decisions, and desired results. We will explore some important design considerations which include: (1) technology factors – sensors, indicators, power, resilience, containment; (2) human factors – movement, accessibility, perception, somatic sensibilities; (3) form factors – aesthetics, socio-cultural acceptance, fixing to the body; and (4) interaction factors – modality, soma design, usability, application.

Read:

Wang, W., Bryan-Kinnis, N., & Yan, Q. (2015). *The design space and the shifting trigger in wearable product development.* 206–210.

11.2.3.4 Week 10 Designing with the Body

Key themes: soma design, soma design theory, soma design methods, soma design evaluation, training somaesthetic skills

Soma design places first-person aesthetic sensory experience and bodily knowing in the center of the design process. Inspired mainly by Richard

Shusterman's theory of somaesthetics supported by Sheets-Johnstone's conceptualizations of kinaesthetics, it engages with the ways in which recognizing sensuous experiences can help us examine the connections between sensations, feelings, affect, and meaning making. Soma design is more of an exploration – through our senses, movements, and material encounters – of the countless possibilities emerging during the design process. A soma design process strives to realize the aesthetic modalities of both the materials and the creative process. One of the main assumptions behind soma-based approaches to design is that our somas not only shape our designs but can also serve as a design material. Although design researchers and practitioners have been trying to establish concrete knowledge and design methods in relation to movement, kinaesthetics, emotional expressions, coenesthesia, proprioception, and tactility, designing for the lived felt body, its affectivity and vivacity has been notably absent from both theory and practice. Most design work engaging the body has been grounded in an instrumental view of both the body and its interactions. Despite the lack of a somehow established model of the soma design process, there is a multiplicity of methods and approaches that have been utilized to explore the potential of the soma as a design modality. We are going to examine these approaches through a consideration of the importance of acquiring somaesthetic skills, engagement in somaesthetic brainstorming, and the consideration of the role of materials in the "dynamic gestalt" of the interaction (Höök, 2018).

We are then going to look at the practical utilization of soma design to movement-based games. These present an interesting case as they do not fit the dominant "button-press game" paradigm and require novel creative approaches to movement interaction.

Read:

Höök, K., Ståhl, A., Jonsson, M., Mercurio, J., Karlsson, A., & Johnson, E.-C. B. (2015). Somaesthetic design. *Interactions, 22*(4), 26–33.

Françoise, J., Candau, Y., Fdili Alaoui, S., & Schiphorst, T. (2017). Designing for Kinesthetic Awareness: Revealing User Experiences through Second-Person Inquiry. *Proceedings of the 2017 CHI Conference on Human Factors in Computing Systems*, 5171–5183.

Mueller, F., & Isbister, K. (2014). Movement-Based Game Guidelines. *Proceedings of the 32nd Annual ACM Conference on Human Factors in Computing Systems – CHI '14*, 2191–2200.

11.2.4 Module III Contemporary Debates

11.2.4.1 Week 11 Ethical Considerations

Key themes: considering the "other" in the design, production, evaluation, and application of sports technologies

The theme of this week explores the ethical implications surrounding the values and epistemologies underlying the design, evaluation, and application of technologies in sports. Building on the specific case of equitation, we are going to discuss how technologies have been utilized within anthropocentric legitimations of violence and a lack of empathy for the "Other."

Equitation refers to activities where a horse interacts with a human. Although not necessarily digital, technologies – such as saddles, bridles, bits, spurs, and whips among many others – have been used for centuries to tame, train, control, and often abuse horses. The abuse of horses happens not only through technologies, but also through techniques and processes that disregard an animal's welfare. Assuming the state of the "other" through mental imaging of being "them," we are going to engage with notions of courtesy, respect, and compassion toward human and non-human co-participants in sports and extend these to design principles, such as empathy-oriented design.

It is important that we emphasize the need for a novel "more-than-human" design paradigm. Such an approach will improve the quality of knowledge that underlies the design and utilization of interactive technologies in sports and serve as a means "of becoming a better human in an ethical sense, able to connect with, empathize with, and care for others" (Loke & Schiphorst, 2018).

The emerging field of interactive sports technologies can serve as a point of departure toward the creation of technologies that facilitate awareness through renditions of self/less consciousness. We can establish self/less consciousness as a new aesthetic dimension of interaction.

Read:

Randle, H., Steenbergen, M., Roberts, K., & Hemmings, A. (2017). The use of the technology in equitation science: A panacea or abductive science? *Applied Animal Behaviour Science, 190*, 57–73.

Smith, S. J. (2015). Dancing with Horses: The Science and Artistry of Coenesthetic Connection. In N. Carr (Ed.), *Domestic animals and leisure* (pp. 216–240). Palgrave Macmillan UK.

11.2.4.2 Week 12 Case Studies – Designing for Reflection

Key themes: slow technology

With the increasingly ubiquitous nature of computing, the use of digital technologies is shifting from specific, time-limited contexts to becoming an integral part of routine activities and requiring our attention for longer periods of time. In many cases, the excess of information to which we get exposed can be overwhelming. This initiates a need for a shift in the dominant interaction design paradigm – preoccupied with efficiency – to a consideration of reflection, solitude, and mental relaxation. The introduction of *slow technologies* presents a potential solution to this problem. These are technologies designed to "reside in the periphery of our attention, continuously providing us with contextual information without demanding a conscious effort on our behalf." (Hallnäs & Redström, 2001, p. 202). The distinction between slow and fast technology is not defined in terms of time, but serves as a metaphor representing time presence. When we use a technology counting on its efficiency as a tool, time disappears – we get things done in the fastest possible way. Accepting an invitation for reflection inherent in the design of a technology means that time will reappear and thus we open ourselves up for time presence. Extending on this idea, we will discuss how technologies can actively promote moments of reflection and mental rest through their design. Some considerations include: using soma-based approaches to design, how can we design for slowness and mental rest? How can we design for slower, less demanding ways of life?

The conceptualization of reflection as a type of affordance embedded in a computational system can be utilized in technologies for sports. By intentionally slowing an athletic performance down, we create a possibility for a reflective space where we can gain critical distance to process the implications of our movements.

Read:

Hallnäs, L., & Redström, J. (2001). Slow technology – Designing for reflection. *Personal and Ubiquitous Computing, 5*(3), 201–212.

11.2.4.3 Week 13 Case Studies – Augmented Realities for Sports

Final week. Open discussion.

Read:

Kajastila, R., & Hämäläinen, P. (2014). Augmented Climbing: Interacting with Projected Graphics on a Climbing Wall. *Proceedings of the Extended Abstracts of the 32nd Annual ACM Conference on Human Factors in Computing Systems – CHI EA '14,* 1279–1284.

11.3 Learning Objectives

11.3.1 Concepts and Theory

1. Create a sense of the historical context, theoretical frameworks, and contemporary debates surrounding the idea of embodiment in relation to skill acquisition in sports and technology.
2. Understand movement and physical activity beyond reductionist functionality (as a means to an end) to include its experiential dimension.
3. Reimagine the future of sports training through an articulation of the epistemological essence of somatically-informed athletic practices facilitated by interactive technologies.
4. Expand the theories and practice of embodiment to the domain of Sports-HCI.
5. Rearticulate theories of embodied design to meet the specific context of interactive technologies for sports.

11.3.2 Skills and Technique

6. Apply emergent embodied approaches to the design of interactive technologies which support the research and practice of skill acquisition in sports.
7. Integrate movement laboration[1] in the form of theoretical and experiential frameworks to body-based technology design, particularly in its use as a design resource for interactive sports technologies.
8. Incorporate somatic and kinaesthetic expertise within the context of athletic performance in order to extend on the practical application of embodied theory to interactive sports technologies.
9. Display the ability to engage in experimental design for athletic performance through the epistemological and methodological frameworks of embodiment.
10. Evaluate technologies for athletic performance through the lens of somatic engagement and somatic sensitivities.

11.3.3 Reflection and Criticism

11. Learn to identify, recognize, and harness the intrinsic value of movement.
12. Articulate the ethical challenges inherent in the utilization of technologies in sports through exploring the corporeally-charged consciousness of being in relation to other (human and non-human) participants in athletic activities.
13. Demonstrate the ability to critically approach theoretical perspectives in relation to embodied approaches to human-technology relations.

11.4 Course Assignments

11.4.1 Video Journals (Weeks 2–5)

Using a smartphone or other easily accessible device, record a weekly video clip (~5 minutes in length) that captures how the literature and key ideas presented in class affect your awareness of your body's being in the world. How you approach this assignment is up to you. You may simply recount your thoughts, ideas, and experiences while doing the dishes. You may elaborate on things you were surprised or frustrated by and document the visceral reactions associated with these. You may explain the soma- and kinaesthetic dimensions of class-related activities. Or, you may record real-life experiences exemplifying the ideas discussed in class. The purpose of these journals is not to produce an aesthetically pleasant video, but to facilitate reflection on your corporeal experiencing of the world.

11.4.2 What? How? Why? (Weeks 6–7)

What? How? Why? is a tool developed by the Design School at Stanford University to help designers reach deeper levels of observation[2]. We are going to use this tool for achieving not only deeper levels of observation, but also deeper levels of reflection. The objective of this exercise is to switch your attention from straight observations to more abstract emotions, visceral reactions, and relational flows at play in the background.

How it works:

1. Divide a sheet of paper into three sections: What?, How?, and Why?
2. Start with specific observations – what is a person of interest doing in a specific situation or a photograph? Write down only objective details without making assumptions.
3. Move to understanding – how is the person doing what they are doing? Does the experience appear to be a positive or a negative one? Use lots of adjectives to catch the flow of the situation.
4. Expand to interpretation – why is the person doing what they are doing in the particular way they are doing it? Make informed guesses about their motivations and emotions.

Now, the assignments:

For week 6, observe a family member or a friend perform a daily task – eating breakfast, drinking water, playing with their kids, folding the laundry, cooking, cleaning the house, driving a car. Analyze the individual's movements and actions, as well as more subjective interpretations of emotional states and behavior using the What? How? Why? framework and quickly document your responses. After doing so, ask your friend/family member what, how, and why they were doing. First, you have to stabilize their attention and then turn it from what to how they were doing. Ask supporting – but not leading – questions to retrospectively access their lived experiences, moving from general representations to a singular experience. Try to navigate them through the various dimensions of their experience and how it was felt through their bodies. Compare their answers with your notes. Were your interpretations and assumptions correct? Document the whole process in detail.

For week 7, you have to engage in a sports/athletic activity. Ask somebody to take a short video of you while performing the task. Before looking at the video footage, document your experience using the What? How? Why? framework. Pay specific attention to your felt senses and the mental images of your body penetrating the space, its surface, and moment-to-moment change. Be as specific as possible. Now, watch the video. Do the what's, how's, and why's of your mental imagery match the factual recording of your moving body?

11.4.3 Quick and Dirty Design (Week 10)

Based on the information you documented in week 7 assignment, you have to plan and design (but not produce) an interactive system which supports – through notions of embodiment – the athletic activity you performed. Provide a detailed explanation of the system's design and how your embodied experiences have informed the design process. Explain how the users are supposed to "feel" what the system aims to deliver. You can also use sketches to visualize your ideas.

11.4.4 Grant Proposal (Weeks 11–13)

Drawing on the literature, key themes, and discussions presented in the course, prepare a grant proposal that targets the **production and evaluation of an interactive sports technology (can be a system, interactive prototype, or other) that explores the junction of embodiment, athletic performance, technology, and computation.** The objective of this assignment is to familiarize you with the experience of (1) planning the creation and evaluation of an interactive sports technology along with relevant design methods and concerns, (2) considering budgeting and costs, (3) writing a research proposal which situates the project within wider theoretical technologies, (4) locating relevant grants, and (5) putting together a grant proposal.

11.4.4.1 Step I – Designing the Study (Weeks 11–12)

This is an iterative assignment – there is a component due each week. After completing a thorough review of the relevant literature and identifying key themes and considerations, there are several stages in the production and evaluation of a system that need to be considered.

1. **Formulate a research question/problem** that you think is important and can be addressed by the utilization of an interactive sports technology. The research question should relate – at least partially – to the theory and practice of embodiment, but the angle/perspective within such exploration is up to you.
2. **Describe the project and how it would be produced.** In this part, you have to consider the design, implementation, testing, and validation of the system. Although there are numerous models that systematize the design process, the User-Centered System Design (UCSD) model (Smith-Atakan, 2006) presents one of the main-stream approaches to the design process:

   1. Task analysis – consideration of what users want.
   2. Requirements gathering – iterative understandings of what and how the system should perform.

3. Prototype implementation – relatively low-cost implementation of a system
4. Design and storyboarding – visualization of the design ideas that users can test.
5. Evaluation – the collection of tangible evidence of how a system is actually used by the users.
6. Installation – the finished system.

We have to keep in mind that somaesthetics-informed tactics to design propose a fundamentally different approach than the most popular design process models today (Höök, 2018). Soma-based design "entails a qualitative shift from a predominantly symbolic, language-oriented stance to an experiential, felt, aesthetic stance permeating the whole design and use cycle" (ibid, p. 175) and as such draws attention to the first-person perspective of the designer and to movement, affect, vital flows, emotions, reflection, experience, and reasoning. Some of the design process strategies proposed by soma design methods include embodied sketching, move to be moved, digital and technological material design encounters (ibid), storytelling (Núñez Pacheco, 2017), Alexander technique, Material for the Spine (MFS) (Candau et al., 2018), and live coding interactions (Françoise et al., 2017), autoethnographic explorations, facilitated interactions, and design probes (Núñez-Pacheco & Loke, 2015) among others. We discussed the soma-design methods in the Sports and HCI module of the course. You are free to select a model that works for your project or mix-and-match models as you see fit. However, make sure that you take into consideration the *embodiment* aspect of your project.

3. Describe in detail how you would collect and document data on and evaluate the success of the project.

Evaluation is one of the final stages of the design process as evidenced in the UCSD model above. It is the process of testing and improving design ideas. During that stage, users (can be members of the design team or external individuals) interact with the system and provide input on its performance. There are many ways in which a system can be evaluated depending on the type of information the system creators want to collect. Some of the more common methods include (but are not limited to) observational evaluation, protocol analysis, experiments, cognitive walkthrough, interviews, and questionnaires. Soma-based design evaluation techniques provide alternative evaluation methods emphasizing the first-person perspective of the users. These include focusing, felt sense (Núñez-Pacheco & Loke, 2018), reflection (Baumer, 2015), second person inquiry through phenomenology (Françoise et al., 2017; Petitmengin, 2006), and micro-phenomenology (Prpa et al., 2020) among others.

11.4.4.2 Step II – Writing the Grant Proposal (Week 13)

Although the formats of grant proposals differ depending on the grant agency's specifications and disciplinary orientation of the research, it usually covers the following elements:

1.   Problem or objective of the proposed study/project

Grant proposals usually begin with an introductory section which exposes the goals and objectives of the project, what is being studied/developed, why it is worth studying/developing, and what the theoretical and/or practical significance of the project is.

2.   Review of the literature/relevant work

The project needs to be situated within the broader context of relevant scholarship or practical work in the area. This establishes your ethos and competency as a researcher/designer and demonstrates that your ideas emerge in a conversation with previous work in the field.

3.   Research question/problem to be addressed

This is a clear formulation of the question/problem around which you center the fundamental essence of your study/project.

4.   Methodology

This section presents a detailed account of how you are going to conduct your study or complete your project. It should include considerations of the study subjects/users, type of measurements, data collection methods, data analysis, ethical or legal implications, schedule, and budget.

5.   Significance of the study/project

In this section, you emphasize why your study/project is important and how it benefits the academic and/or broader community.

Notes

1 Laboration is a Swedish word referring to experimental work in a laboratory. Cheryl Akner-Koler utilizes it in a figurative sense to connote an interactive, embodied experiment performed together with others. It aims at inducing an aesthetic reactions and reasoning, creating a platform for playful exchanges and vitality (Akner Koler, 2007).

2 The exercise and instructions are adopted from the Design School at Stanford University. The document can be found at: https://static1.squarespace.com/static/57c6b79629687fde090a0fdd/t/5b19b2f2aa4a99e99b26b6bb/1528410876119/dschool_bootleg_deck_2018_final_sm+%282%29.pdf.

References

Akner Koler, C. (2007). *Form & formlessness: Questioning aesthetic abstractions through art projects, cross-disciplinary studies and product design education.* Chalmers Univ. of Technology.

Baumer, E. P. S. (2015). Reflective Informatics: Conceptual Dimensions for Designing Technologies of Reflection. In *Proceedings of the 33rd Annual ACM Conference on Human Factors in Computing Systems – CHI '15*. 585–594. New York, NY, United States: Association for Computing Machinery. 10.1145/2702123.2702234

Bødker, S. (2006). When Second Wave HCI Meets Third Wave Challenges. In *Proceedings of the 4th Nordic Conference on Human-Computer Interaction: Changing Roles.* 1–8. New York, NY, United States: Association for Computing Machinery. 10.1145/1182475.1182476

Candau, Y., Schiphorst, T., & Françoise, J. (2018). Designing from Embodied Knowing: Practice-Based Research at the Intersection Between Embodied Interaction and Somatics. In M. Filimowicz & V. Tzankova (Eds.), *New directions in third wave human-computer interaction: Volume 2—Methodologies* (pp. 203–230). Springer International Publishing. 10.1007/978-3-319-73374-6_11

Dourish, P. (2001). *Where the action is: The foundations of embodied interaction.* MIT Press.

Dourish, P. (2004). What we talk about when we talk about context. *Personal and Ubiquitous Computing, 8*(1), 19–30. 10.1007/s00779-003-0253-8

Françoise, J., Candau, Y., Fdili Alaoui, S., & Schiphorst, T. (2017). Designing for Kinesthetic Awareness: Revealing User Experiences through Second-Person Inquiry. In *Proceedings of the2017 CHI Conference on Human Factors in Computing Systems.* 5171–5183. New York, NY, United States: Association for Computing Machinery. 10.1145/3025453.3025714

Hallnäs, L., & Redström, J. (2001). Slow Technology – Designing for Reflection. *Personal and Ubiquitous Computing, 5*(3), 201–212. 10.1007/PL00000019

Höök, K. (2018). *Designing with the body: Somaesthetic interaction design.* The MIT Press.

Loke, L., & Schiphorst, T. (2018). The somatic turn in human-computer interaction. *Interactions, 25*(5), 54–58. 10.1145/3236675

Moen, J. (2005). Towards People Based Movement Interaction and Kinaesthetic Interaction Experiences. 121. New York, NY, United States: Association for Computing Machinery. 10.1145/1094562.1094579

Mueller, F., & Young, D. (2018). 10 Lenses to Design Sports-HCI. *Foundations and Trends® in Human–Computer Interaction, 12*(3), 172–237. 10.1561/1100000076

Núñez Pacheco, C. (2017). *Designing for aesthetic experiences from the body and felt-sense.* University of Sydney.

Núñez-Pacheco, C., & Loke, L. (2015). The Felt Sense Project: Towards a Methodological Framework for Designing and Crafting From the Inner Self. In

Proceedings of the 21st International Symposium on Electronic Art. Vancouver, BC, Canada: ISEA.

Núñez-Pacheco, C., & Loke, L. (2018). Towards a technique for articulating aesthetic experiences in design using focusing and the felt sense. *The Design Journal, 21*(4), 583–603. 10.1080/14606925.2018.1467680

Nylander, S., Tholander, J., Mueller, F., & Marshall, J. (2014). HCI and Sports. In *Proceedings of the Extended Abstracts of the 32nd Annual ACM Conference on Human Factors in Computing Systems - CHI EA '*. 14, 115–118. New York, NY, United States: Association for Computing Machinery. 10.1145/2559206.2559223.

Petitmengin, C. (2006). Describing one's subjective experience in the second person: An interview method for the science of consciousness. *Phenomenology and the Cognitive Sciences, 5*(3–4), 229–269. 10.1007/s11097-006-9022-2

Polanyi, M. (1966). *The tacit dimension.* Doubleday & Company.

Prpa, M., Fdili-Aloui, S., Schiphorst, T., & Pasquier, P. (2020). Articulating Experience: Reflections from Experts Applying Micro-Phenomenology to Design Research in HCI. In *Proceedings of the Conference on Human Factors in Computing Systems. CHI'20.* HI, USA.

Schiphorst, T. (2007). Really, really small: The palpability of the invisible. In *Proceedings of the 6th ACM SIGCHI Conference on Creativity & Cognition – C&C '07*, 7–16. New York, NY, United States: Association for Computing Machinery. 10.1145/1254960.1254962

Schön, D. A. (1983). *The reflective practitioner: How professionals think in action.* Basic Books.

Sheets-Johnstone, M. (2011). *The primacy of movement* (Expanded 2nd ed). John Benjamins Publishing.

Sheets-Johnstone, M. (2014). Animation: Analyses, elaborations, and implications. *Husserl Studies, 30*(3), 247–268. 10.1007/s10743-014-9156-y

Shusterman, R. (2008). *Body consciousness: A philosophy of mindfulness and somaesthetics.* Cambridge University Press.

Smith-Atakan, S. (2006). *Human-computer interaction.* Thomson.

Whitehead, M. (Ed.). (2010). *Physical literacy: Throughout the lifecourse* (1st ed). Routledge.

Index

Note: Page numbers followed by "n" refer to notes; page numbers in **Bold** refer to tables; and page numbers in *italics* refer to figures

Printed in the United States
by Baker & Taylor Publisher Services